T0250134

Mechanics of Project Management

Nuts and Bolts of Project Execution

Analytics and Control

Series Editor

Adedeji B. Badiru

Air Force Institute of Technology, Dayton, Ohio

Mechanics of Project Management

Nuts and Bolts of Project Execution

By
Adedeji B. Badiru
S. Abidemi Badiru
I. Adetokunboh Badiru

CRC Press
Taylor & Francis Group
Boca Raton London New York

CRC Press is an imprint of the
Taylor & Francis Group, an **informa** business

CRC Press
Taylor & Francis Group
6000 Broken Sound Parkway NW, Suite 300
Boca Raton, FL 33487-2742

© 2019 by Taylor & Francis Group, LLC
CRC Press is an imprint of Taylor & Francis Group, an Informa business

No claim to original U.S. Government works

Printed on acid-free paper

International Standard Book Number-13: 978-1-1383-4882-0 (Hardback)

Visit the Taylor & Francis Web site at
http://www.taylorandfrancis.com

and the CRC Press Web site at
http://www.crcpress.com

*To Blake and Alexa, the project managers
of the future*

Contents

Preface

*M*ECHANICS OF *P*ROJECT *M*ANAGEMENT offers a unique guide to the management of projects, both big and small, in all spheres of human endeavor. The book presents the nuts and bolts of untangling the typical knots in project execution. It is applicable to all types of projects, including technology installation, manufacturing, financial endeavors, science projects, engineering implementations, and personal projects. This book presents a step-by-step application of project management techniques to manage any project. The book uses the Project Management Body of Knowledge (PMBOK®) framework from the Project Management Institute (PMI) as the platform for the topics covered, coupled with a systems-engineering structure. Technical project management is the basis for sustainable national advancement, which often depends on the development of corporate projects. As such, managing corporate projects effectively is essential for national economic vitality.

Project management is the process of managing, allocating, and timing resources to achieve a given goal in an efficient and expeditious manner. Project management techniques are widely seen in many endeavors, including food services, financial services, manufacturing enterprises, construction management, marketing, health care delivery, transportation, defense, and public services.

Project management represents an excellent basis for integrating various management techniques such as statistics, operations research, Six Sigma, computer simulation, and so on, within business and industry. The purpose of this book is to present a focused approach to project management for all readers. The integrated approach covers the concepts, tools, and techniques of project management. The elements of the Project Management Body of Knowledge provide a unifying platform for the topics covered in the book, which is useful as a supplementary reading for practicing engineers and as a guidebook for field operators. It will appeal to all professionals because of its focused treatment of the topics. *Mechanics of Project Management* provides a manifestation of practical management of all types of projects.

Acknowledgments

We thank our immediate and extended family members, who continue to recognize and appreciate the value of fellowship and togetherness from the perspective of effective project execution.

Authors

ADEDEJI B. BADIRU is a professor of Systems Engineering and dean of the Graduate School of Engineering and Management at the Air Force Institute of Technology (AFIT). He was previously professor and head of Systems Engineering and Management at AFIT, professor and department head of Industrial & Information Engineering at the University of Tennessee in Knoxville, and professor of Industrial Engineering and Dean of University College at the University of Oklahoma, Norman. He is a registered professional engineer (PE), a certified project management professional (PMP), a fellow of the Institute of Industrial Engineers, and a fellow of the Nigerian Academy of Engineering. He holds BS in Industrial Engineering, MS in Mathematics, and MS in Industrial Engineering from Tennessee Technological University, and PhD in Industrial Engineering from the University of Central Florida. His areas of interest include mathematical modeling, project control economic analysis, technology transfer systems engineering modeling, and innovation strategies. He is the author of several books and technical journal articles. He is a member of several professional associations and scholastic honor societies.

S. Abidemi Badiru, MBA, PMP, is a process improvement engineer for a multinational financial services company in the Dallas, Texas area. She has a BS in Chemical Engineering from the University of Oklahoma and MBA from the University of Texas at Dallas. She has over 19 years of previous senior-level experience in a variety of food service companies. She is a registered project management professional. She has conducted several process improvement projects around the nation and is a co-author of *Industrial Project Management*, published by CRC Press.

I. Adetokunboh Badiru, MS, is a senior engineer at General Motors in Detroit, Michigan. He has a BS in Mechanical Engineering from the University of Oklahoma and MS from the University of Michigan, Ann Arbor. He has two USA patents awarded and seven more pending. His technical expertise and responsibilities at GM are currently in the area of Features Owner (Steering). He was previously a lead engineer for Steering Performance for GM Premium Luxury Vehicles. He has published several technical journal articles and is a co-author of *Industrial Project Management*, published by CRC Press.

CHAPTER 1

Project Integration

This focus book is a concise presentation using the traditional framework of the project management body of knowledge (PMBOK®) espoused by the Project Management Institute (PMI). The conciseness of the book is achieved by focusing on the core mechanics of each element in the framework. The main knowledge areas of PMBOK are listed below (PMI, 2017):

1. Integration
2. Scope Management
3. Time Management
4. Cost Management
5. Quality Management
6. Human Resource Management
7. Communication Management
8. Risk Management
9. Procure Management

This book presents a concise chapter on each of the knowledge areas. Thus, this first chapter is on project integration. The elements of project integration include the following:

- Integrative project charter

- Project scope statement

- Project management plan

- Project execution management

- Change control

The framework for this book is predicated on presenting guiding lessons and establishing curiosity for more study of project management tools and techniques.

> "Lesson learned should be lesson practiced."
>
> *ADEDEJI BADIRU, 2012*

> "When curiosity is established, the urge to learn develops."
>
> *ADEDEJI BADIRU, 2013*

WHAT IS INTEGRATION?

In the simplest of terms, integration means bringing everything together. The rationale for making integration the first area in the project management body of knowledge is the need to recognize and account for all the nuances that are downstream in project execution. If an element is recognized at the integration stage, it is less likely to be overlooked downstream in the project. For the purpose of enunciating integration in this focus book, we will use the DEJI® systems model (Badiru, 2014) as the guiding framework. Figure 1.1 presents the model. The model advocates taking each

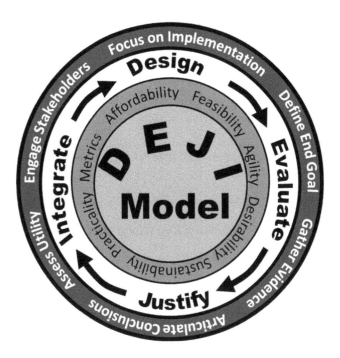

FIGURE 1.1   DEJI® systems model for project integration.

process through the stages of Design, Evaluation, Justification, and Integration. Design, in this context, can represent process design, concept formulation, plan generation, group decision, and so on. This stage then leads to the next stage of evaluation, in which whatever is generated in the design stage is rigorously evaluated with respect to strengths, weaknesses, opportunities for enhancement, and threats to implementation. This stage-to-stage systems process has been effectively applied to quality management, product development, and educational curriculum development. Within the model's structure, sub-topics addressed include metrics, such as feasibility, affordability, desirability, reliability, practicality, and sustainability. The outer shell of the model covers focus on the end goal, implementation strategies, stakeholder engagement, utility assessment, data analytics, and articulation of decisions and conclusions. This sequential process is suitable for the application of the ICOM (Input, Controls, Outcomes, and Mechanisms) process, shown in Figure 1.2. Thus, as a project moves along, each stage should take account of the input, process, and outputs elements of that stage. The following literary laws are applicable to any project environment and provision must be made to mitigate their adverse impacts:

Parkinson's Law: Work expands to fill the available time or space.

Peter's Principle: People rise to the level of their incompetence.

Murphy's Law: Whatever can go wrong will.

Badiru's Rule: The grass is always greener where you most need it to be dead.

The elements in the ICOM Model are explained in the following paragraphs.

Input

Project activities start with the customer stating the requirements of interest. This serves as the input to the process. The inputs from

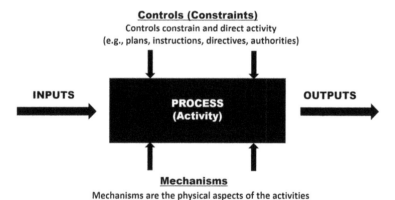

FIGURE 1.2   ICOM process for project management.

the customers represent the objectives of the activities in question, thereby guiding the scope of work.

Constraints

All activities are subject to some sort of constraints and controls. Constraints always exist and cannot be totally removed or avoided. What is needed is to mitigate any operational impediments associated with the constraints. Some of the constraints are financial (budget), time (deadlines), legal, environmental, and/or quality. If a constraint is recognized earlier on, the project will be better prepared to deal with it. Further, constraints may dictate the need to adjust the objectives. There are often compromises between the "wants" and "needs" of the customer versus the capabilities of performing organization.

Mechanism

A mechanism is the specific way of accomplishing the objectives. This is where the "mechanics" part of the title of this book comes into play. The tools and techniques of project execution relate to "how" to pursue an objective. Mechanisms move a project forward. The mechanisms are a combination of people, knowledge, expertise, talents, capital, policies, procedures, authorities, and technology.

Output

The output stage of ICOM specifies what will be achieved in the context of the specified objectives, needs, wants, and so on. It is essential that we confirm if the outputs satisfy the expectations of the customer. Overall, ICOM provides a clear direction on what critical success areas to consider. It also guides the process of identifying key result areas (KRA) and key performance indicators (KPI).

Project integration management specifies how the various parts of a project come together to make up the complete project. This knowledge area recognizes the importance of linking several aspects of a project into an integrated whole. The Henry Ford quote at the beginning of this chapter emphasizes the importance of "togetherness" in any project environment. Project integration management area includes the processes and activities needed to identify, define, combine, unify, and coordinate the various processes and project activities. The traditional concepts of systems analysis are applicable to project processes. The definition of a project system and its components refers to the collection of interrelated elements organized for the purpose of achieving a common goal. The elements are organized to work synergistically together to generate a unified output that is greater than the sum of the individual outputs of the components. The harmony of project integration is evident in the characteristic symbol that this book uses to denote this area of project management knowledge.

While the knowledge areas of project management, as discussed in Chapter 1, overlap and can be implemented in alternate orders, it is still apparent that Project Integration Management is the first step of the project effort. This is particularly based on the fact that the project charter and the project scope statement are developed under the project integration process. In order to achieve a complete and unified execution of a project, both qualitative and quantitative skills must come into play. Similarly, a hard skill can be as easy as it is difficult. In fact, human resource management issues are often the most difficult to handle; yet they are typically classified as falling under the banner of soft skills. Quantitative modeling skills (hard skills), on the other hand, can be easy once all the input parameters are known and accounted for in the modeling process. Thereafter, the utilization of the model becomes an easy repetition of a proven model. By contrast, the nuances of human emotion, sentiments, and psychological variability make qualitative management not easy at all.

PROJECT INTEGRATION: STEP-BY-STEP IMPLEMENTATION

The seven elements in the block diagram are carried out across the process groups recommended in PMI's PMBOK. The overlay of the elements and the process groups are shown in Table 1.1. Thus, under the knowledge area of Integration, the required steps are:

Step 1: Develop Project Charter

Step 2: Develop preliminary project scope

Step 3: Develop project management plan

Step 4: Direct and manage project execution

Step 5: Monitor and control project work

Step 6: Perform integrated change control

Step 7: Close project

TABLE 1.1   Tools and Techniques for Project Integration Management

	Initiating	Planning	Executing	Monitoring and Controlling	Closing
Project integra-tion	1. Develop project charter	3. Develop project charter manage-ment plan	4. Direct and manage project execu-tion	5. Monitor project work	7. Close project
	2. Develop preliminary project scope			6. Integrated change control	

In addition to the standard PMBOK inputs, tools, techniques, and outputs, the project team may add in-house items of interest to the steps in each process group. Such in-house items are summarized below:

- Inputs: Other in-house (custom) factors of relevance and interest

- Tools and Techniques: Other in-house (custom) tools and techniques

- Outputs: Other in-house outputs, reports, and data inferences of interest to the organization

Figure 1.3 shows the nuts and bolts of project execution. For each step (presented as textual listing) in the project execution sequence, the inputs are presented. Then, the applicable tools and techniques are listed. Finally, the expected outputs are described.

Table 1.2 shows the input-to-output items for developing project charter. The tabular format is useful for explicitly identifying what the project analyst needs to do or use for each step of the project management process. Tables 1.3 through 1.8 present the input-to-output entries for the other steps under Integration Management.

FIGURE 1.3   Nuts and bolts in project execution.

TABLE 1.2   Tools and Techniques for Developing Project Charter within Integration Management

STEP 1: Develop Project Charter

Inputs
Project contract (if applicable)
Project statement of work
Enterprise environmental factors
Organizational process assets
Other in-house (custom) factors of relevance and interest

Tools and Techniques
Project selection methods
AHP (Analytic Hierarchy Process)
Project management methodology
Project management information system
Expert judgment
Balance scorecard
Process control charts
Other in-house (custom) tools and techniques

Output(s)
Project charter
Other in-house outputs, reports, and data inferences of interest to the organization

STEP 1: DEVELOPING PROJECT CHARTER

A project charter formally authorizes a project. It is a document that provides authority to the project manager and it is usually issued by a project initiator or sponsor external to the project organization. The purpose of a charter is to define at a high level what the project is about, what the project will deliver, what resources are needed, what resources are available, and how the project is justified. The charter also represents an organizational commitment to dedicate the time and resources to the project. The charter should be shared with all stakeholders as a part of the communication requirement of the Triple C approach.

Cooperating stakeholders will not only sign-off on the project, but also make personal pledges to support the project. Projects are usually chartered by an enterprise, a government agency, a company, a program organization, or a portfolio organization in response to one or more of the following business opportunities or organizational problems:

- Market demand

- Response to regulatory development

- Customer request

- Business need

- Exploitation of technological advance

- Legal requirement

- Social need

The driving force for a project charter is the need for an organization to make a decision about which projects to authorize to respond to operational threats or opportunities. It is desired for a charter to be brief. Depending on the size and complexity of a project, the charter should not be more than two to three pages. Where additional details are warranted, the expatiating details can be provided as addenda to the basic charter document. The longer the basic charter, the less the likelihood that everyone will read and imbibe the contents. So, brevity and conciseness are desired virtues of good project charters. The charter should succinctly establish the purpose of the project, the participants, and general vision for the project.

The project charter is used as the basis for developing project plans. While it is developed at the outset of a project, a charter should always be fluid. It should be reviewed and updated throughout the life of the project. The components of the project charter are summarized below:

- Project overview
- Assigned project manager and authority level
- Project requirements
- Business needs
- Project purpose, justification, and goals
- Impact statement
- Constraints (time, cost, performance)
- Assumptions
- Project Scope
- Financial implications
- Project approach (policies, procedures)
- Project organization
- Participating organizations and their respective roles and level of participation
- Summary milestone schedule
- Stakeholder influences
- Assumptions and constraints (organizational, environmental, external)
- Business plan and expected return on investment (ROI), if applicable
- Summary budget

The project charter does not include the project plan. Planning documents, which may include a project schedule, quality plan, staff plan, communication hierarchy, financial plan, and risk plan, should be prepared and disseminated separately from the charter.

- **Project Overview**

  - The project overview provides a brief summary of the entire project charter. It may provide a brief history of the events that led to the project, an explanation of why the project was initiated, a description of project intent and the identity of the original project owner.

- **Project Goals**

  - Project goals identify the most significant reasons for performing a project. Goals should describe improvements the project is expected to accomplish along with who will benefit from these improvements. This section should explain what various benefactors will be able to accomplish due to the project. Note that the Triple C approach requires these details as a required step to securing cooperation.

- **Impact Statement**

  - The impact statement identifies the influence the project may have on the business, operations, schedule, other projects, current technology, and existing applications. While these topics are beyond the domain of this project, each of these items should be raised for possible action.

- **Constraints and Assumptions**

  - Constraints and assumptions identify any deliberate or implied limitations or restrictions placed on the project along with any current or future environment the project must accommodate. These factors will influence many project decisions and strategies. The potential impact of each constraint or assumptions should be identified.

- **Project Scope**

  - Project scope defines the operational boundaries for the project. Specific scope components are the areas or functions to be impacted by the project and the work that will be performed. The Project Scope should identify both what is within the scope of the project and what is outside the scope of the project.

- **Project Objectives**

  - Project objectives identify expected deliverables from the project and the criteria that must be satisfied before the project is considered complete.

- **Financial Summary**

  - The financial summary provides a recap of expected costs and benefits due to the project. These factors should be more fully defined in the cost-benefit analysis of the project. Project financials must be reforecast during the life of the effort.

- **Project Approach**

  - Project approach identifies the general strategy for completing the project and explains any methods or processes, particularly policies and procedures that will be used during the project.

- **Project Organization**

  - Project organization identifies the roles and responsibilities needed to create a meaningful and responsive structure that enables the project to be successful. Project organization must identify the people who will play each assigned role. At a minimum, this section should identify who plays the roles of project owner, project manager, and core project team.

  - A project owner is required for each project.

    - This role must be filled by one or more individuals who are the fiscal trustee(s) for the project to the larger organization. This person considers the global impact of the project and deems it worthy of the required expenditure of money and time. The project owner communicates the vision for the effort and certifies the initial project charter and project plan. Should changes be required, the Project owner confirms these changes and any influence on the project charter and project plan. When project decisions cannot be made at the team level, the project owner must resolve these issues. The project owner must play an active role throughout the project, especially ensuring that needed resources have been committed to the project and remain available.

- A Project Manager is required for each project.

  - The project manager is responsible for initiating, planning, executing and controlling the total project effort. Members of the project team report to the project manager for project assignments and are accountable to the project manager for the completion of their assigned work.

DEFINITIONS OF INPUTS TO STEP 1

Contract: A contract is a contractual agreement between the organization performing the project and the organization requesting the project. It is treated as an input if the project is being done for an external customer.

Project statement of work (SOW): This is a narrative description of products or services to be supplied by the project. For internal projects, it is provided by the project initiator or sponsor. For external projects, it is provided by the customer as part of the bid document. For example, request for proposal, request for information, request for bid, or contract statements may contain specific work to be done. The SOW indicates the following:

- Business need based on required training, market demand, technological advancement, legal requirement, government regulations, industry standards, or trade consensus

- Product scope description, which documents project requirements and characteristics of the product or service that the project will deliver

- Strategic plan, which ensures that the project supports the organization's strategic goals and business tactical actions.

Enterprise environmental factors: These are factors that impinge upon the business environment of the organization. They include organizational structure, business culture, governmental standards, industry requirements, quality standards, trade agreements, physical infrastructure, technical assets, proprietary information, existing human resources, personnel administration, internal work authorization system, marketplace profile, competition, stakeholder requirements, stakeholder risk tolerance levels, commercial obligations, access to standardized cost estimating data, industry risk, technology variances, product lifecycle, and project management information systems.

Organizational process assets: These refer to the business processes used within an organization. They include standard processes, guidelines, policies, procedures, operational templates, criteria for customizing standards to specific project requirements, organization communication matrix, responsibility matrix, project closure guidelines (e.g., sunset clause), financial controls procedure, defect management procedures, change control procedures, risk control procedures, process for issuing work authorizations, processes for approving work authorizations, management of corporate knowledge base, and so on.

DEFINITIONS OF TOOLS AND TECHNIQUES FOR STEP 1

Project selection methods: These methods are used to determine which projects an organization will select for implementation. The methods can range from basic seat-of-the-pants heuristics to highly complex analytical techniques. Some examples are benefit measurement methods, comparative measure of worth analysis, scoring models, benefit contribution, capital rationing approaches, budget allocation methods, and graphical analysis tools. Analytical techniques are mathematical models that use linear programming, nonlinear programming, dynamic programming, integer programming, multi-attribute optimization, and other algorithmic tools.

Project management methodology: This defines the set of project management process groups, their collateral processes, and related control functions that are combined for implementation for a particular project. The methodology may or may not follow a project management standard. It may be an adaptation of an existing project implementation template. It can also be a formal mature process or informal technique that aids in effectively developing a project charter.

Project management information system (PMIS): This is a standardized set of automated tools available within the organization and integrated into a system for the purpose of supporting the generation of a project charter, facilitating feedback as the charter is refined, controlling changes to the project charter, or releasing the approved document.

Expert judgment: This is often used to assess the inputs needed to develop the project charter. Expert judgment is available from sources such as the experiential database of the organization, knowledge repository, knowledge management practices, knowledge transfer protocol, business units within the organization, consultants, stakeholders, customers, sponsors, professional organizations, technical associations, and industry groups.

DEFINITION OF OUTPUT OF STEP 1

Project charter: As defined earlier in this chapter, a project charter is a formal document that authorizes a project. It provides authority to the project manager and it is usually issued by a project initiator or sponsor external to the project organization. It empowers the project team to carry out the actions needed to accomplish the end goal of the project.

TABLE 1.3   Tools and Techniques for Developing Preliminary Project Scope
Statement within Integration Management

STEP 2: Develop Preliminary Project Scope Statement

Inputs
Project charter
Project statement of work
Enterprise environmental factors
Organizational process assets
Other in-house (custom) factors of relevance and interest

Tools and Techniques
Project management methodology
Project management information system
Expert judgment
CMMI (Capability Maturity Model Integration)
Critical chain
Process control charts
Other in-house (custom) tools and techniques

Output(s)
Project scope statement

STEP 2: DEVELOP PRELIMINARY PROJECT SCOPE STATEMENT

Project scope presents a definition of what needs to be done. It specifies the characteristics and boundaries of the project and its associated products and services, as well as the methods of acceptance and scope control. Scope is developed based on information provided by the project initiator or sponsor. A scope statement includes the following:

- Project and product objectives

- Product characteristics

- Service requirements

- Product acceptance criteria

- Project constraints

- Project assumptions

- Initial project organization

- Initial defined risks

- Schedule milestones

- Initial work breakdown structure (WBS)

- Order-of-magnitude cost estimate

- Project configuration management requirements

- Approval requirements.

DEFINITIONS OF INPUTS TO STEP 2

Inputs for STEP 2 are the same as defined for STEP 1 covering the project charter, statement of work, environmental factors, and organizational process assets.

DEFINITIONS OF TOOLS AND TECHNIQUES FOR STEP 2

The tools and techniques for STEP 2 are the same as defined for STEP 1 and cover project management methodology, project management information system, and expert judgment.

DEFINITION OF OUTPUT OF STEP 2

The output of STEP 2 is the preliminary project scope statement, which was defined and described earlier.

TABLE 1.4   Tools and Techniques for Developing Project Management Plan
Within Integration Management

STEP 3: Develop Project Management Plan

Inputs
Preliminary project scope statement
Project management processes
Enterprise environmental factors
Organizational process assets
Other in-house (custom) factors of relevance and interest

Tools and Techniques
Project management methodology
Project management information system
Expert judgment

Output(s)
Project management plan

STEP 3: DEVELOP PROJECT MANAGEMENT PLAN

A project management plan includes all actions necessary to define, integrate, and coordinate all subsidiary and complementing plans into a cohesive project management plan. It defines how the project is executed, monitored and controlled, and closed. The project management plan is updated and revised through the integrated change control process. In addition, the process of developing project management plan documents the collection of outputs of planning processes and includes the following:

- Project management processes selected by the project management team

- Level of implementation of each selected process

- Descriptions of tools and techniques to be used for accomplishing those processes

- How selected processes will be used to manage the specific project

- How work will be executed to accomplish the project objectives

- How changes will be monitored and controlled

- How configuration management will be performed

- How integrity of the performance measurement baselines will be maintained and used

- The requirements and techniques for communication among stakeholders

- The selected project lifecycle and, for multi-phase projects, the associated project phases

- Key management reviews for content, extent, and timing

The project management plan can be a summary or integration of relevant subsidiary, auxiliary, and ancillary project plans. All efforts that are expected to contribute to the project goal can be linked into the overall project plan, each with the appropriate level of detail. Examples of subsidiary plans are the following:

- Project scope management plan

- Schedule management plan

- Cost management plan

- Quality management plan

- Process improvement plan

- Staffing management plan

- Communication management plan

- Risk management plan

- Procurement management plan

- Milestone list

- Resource calendar

- Cost baseline

- Quality baseline

- Risk register

DEFINITION OF INPUTS TO STEP 3

Inputs to STEP 3 are the same as defined previously and include the preliminary project scope statement, project management processes, enterprise environmental factors, and organizational process assets.

DEFINITION OF TOOLS AND TECHNIQUES FOR STEP 3

The tools and techniques for STEP 3 are project management methodology, project information system, and expert judgment. Project management methodology defines a process which aids a project management team in developing and controlling changes to the project plan. The project management information system at this step covers the following segments:

- Automated system, which is used by the project team to do the following:

  - Support generation of the project management plan

  - Facilitate feedback as the document is developed

  - Control changes to the project management plan

- Release the approved document
- Configuration management system, which is a subsystem that includes sub-processes for accomplishing the following:
  - Submitting proposed changes
  - Tracking systems for reviewing and authorizing changes
  - Providing a method to validate approved changes
  - Implementing change management system
- Configuration management system, which forms a collection of formal procedures used to apply technical and administrative oversight to do the following:
  - Identify and document functional and physical characteristics of a product or component
  - Control any changes to such characteristics
  - Record and report each change and its implementation status
  - Support audit of the products or components to verify conformance to requirements
- The change control system is the segment of project management information system that provides a collection of formal procedures that define how project deliverables and documentation are controlled.

Expert judgment, the third tool for STEP 3, is applied to develop technical and management details to be included in the project management plan.

DEFINITION OF OUTPUT OF STEP 3

The output of STEP 3 is the project management plan.

TABLE 1.5   Tools and Techniques for Managing Project Execution Within
Integration Management

STEP 4: Direct and Manage Project Execution

Inputs
Project management plan
Approved corrective actions
Approved preventive actions
Approved change requests
Approved defect repair
Validated defect repair
Administrative closure procedure
Other in-house (custom) factors of relevance and interest

Tools and Techniques
Project management methodology
Project management information system
Process flow diagram
Other in-house (custom) tools and techniques

Output(s)
Project execution plan

STEP 4: DIRECT AND MANAGE PROJECT EXECUTION

STEP 4 requires the project manager and project team to perform
multiple actions to execute the project plan successfully. Some
of the required activities for project execution are summarized
below:

- Perform activities to accomplish project objectives

- Expend effort and spend funds

- Staff, train, and manage project team members

- Obtain quotation, bids, offers, or proposals as appropriate

- Implement planned methods and standards

- Create, control, verify, and validate project deliverables

- Manage risks and implement risk response activities

- Manage sellers

- Adapt approved changes into scope, plans, and environment

- Establish and manage external and internal communication channels

- Collect project data and report cost, schedule, technical`, and quality progress and status information to facilitate forecasting

- Collect and document lessons learned and implement approved process improvement activities

The process of directing and managing project execution also requires implementation of the following:

- Approved corrective actions that will bring anticipated project performance into compliance with the plan

- Approved preventive actions to reduce the probability of potential negative consequences

- Approved defect repair requests to correct product defects during quality process

DEFINITION OF INPUTS TO STEP 4

Inputs to STEP 3 are summarized as follows:

- Project management plan

- Approved corrective actions: These are documented, authorized directions required to bring expected future project performance into conformance with the project management plan.

- Approved change requests: These include documented, authorized changes to expand or contract project scope. Can also modify policies, project management plans, procedures, costs, budgets, or revise schedules. Change requests are implemented by the project team.

- Approved defect repair: This is documented, authorized request for product correction of defect found during the quality inspection or the audit process.

- Validated defect repair: This is a notification that re-inspected repaired items have either been accepted or rejected.

- Administrative closure procedure: Documents all activities, interactions, and related roles and responsibilities needed in executing the administrative closure procedure for the project.

DEFINITIONS OF TOOLS AND TECHNIQUES FOR STEP 4

The tools and techniques for STEP 4 are project management methodology and project management information system and they were previously defined.

DEFINITIONS OF OUTPUTS OF STEP 4

- Deliverables

- Requested changes

- Implemented change requests

- Implemented corrective actions

- Implemented preventive actions

- Implemented defect repair

- Work performance information

TABLE 1.6   Tools and Techniques for Monitoring and Controlling Project Work Within Integration Management

---

STEP 5: Monitor and Control Project Work

Inputs
Project management plan
Work performance information
Rejected change requests
Other in-house (custom) factors of relevance and interest

Tools and Techniques
Project management methodology
Project management information system
Earned value management
Expert judgment
Other in-house (custom) tools and techniques

Output(s)
Recommended corrective actions
Recommended preventive actions
Forecasts
Recommended defect repair
Requested changes
Other in-house outputs, reports, and data inferences of interest to the
  organization

STEP 5: MONITOR AND CONTROL PROJECT WORK

No organization can be strategic without being quantitative. It is through quantitative measures that a project can be tracked, measured, assessed, and controlled. The need for monitoring and control can be evident in the Request for Quantification (RFQ) that some project funding agencies use. Some quantifiable performance measures are schedule outcome, cost effectiveness, response time, number of reworks, and lines of computer codes developed. Monitoring and controlling are performed to monitor project processes associated with initiating, planning, executing, and closing and is concerned with the following:

- Comparing actual performance against plan

- Assessing performance to determine whether corrective or preventive actions are required, and then recommending those actions as necessary

- Analyzing, tracking, and monitoring project risks to make sure risks are identified, status is reported, response plans are being executed

- Maintaining an accurate timely information base concerning the project's products and associated documentation

- Providing information to support status reporting, progress measurement, and forecasting

- Providing forecasts to update current cost and schedule information

- Monitoring implementation of approved changes

DEFINITION OF INPUTS TO STEP 5

Inputs to STEP 5 include the following:

- Project management plan

- Work performance plan

- Rejected change requests

  - Change requests

  - Supporting documentation

  - Change review status showing the disposition of rejected change requests

DEFINITION OF TOOLS AND TECHNIQUES FOR STEP 5

- Project management methodology.

- Project management information system.

- Earned value technique: This measures performance as the project moves from initiation through closure. It provides means to forecast future performance based on past performance.

- Expert judgment.

DEFINITION OF OUTPUTS OF STEP 5

- Recommended corrective actions: Documented recommendations required to bring expected future project performance into conformance with the project management plan.

- Recommended preventive actions: Documented recommendations that reduce the probability of negative consequences associated with project risks.

- Forecasts: Estimates or predictions of conditions and events in the project's future based on information available at the time of the forecast.

- Recommended defect repair: Some defects found during the quality inspection and audit process recommended for correction.

- Requested changes.

TABLE 1.7   Tools and Techniques for Integrated Change Control within
Integration Management

STEP 6: Perform Integrated Change Control

Inputs

Project management plan

Requested changes

Work performance information

Recommended preventive actions

Recommended corrective actions

Deliverables

Other in-house (custom) factors of relevance and interest

Tools and Techniques

Project management methodology

Project management information system

Expert judgment

Other in-house (custom) tools and techniques

Output(s)

Approved change requests

Rejected change requests

Update project management plan

Update project scope statement

Approved corrective actions

Approved preventive actions

Approved defect repair

Validated defect repair

Deliverables

Other in-house outputs, reports, and data inferences of interest to the
  organization

STEP 6: INTEGRATED CHANGE CONTROL

Integrated change control is performed from project inception
through completion. It is required because projects rarely run
according to plan. Major components of integrated change control include the following:

- Identifying when a change needs to occur or when a change has occurred

- Amending factors that circumvent change control procedures

- Reviewing and approving requested changes

- Managing and regulating flow of approved changes

- Maintaining and approving recommended corrective and preventive actions

- Controlling and updating scope, cost, budget, schedule, and quality requirements based upon approved changes

- Documenting the complete impact of requested changes

- Validating defect repair

- Controlling project quality to standards based on quality reports

Combining configuration management system with integrated change control includes identifying, documenting, and controlling changes to the baseline. Project-wide application of the configuration management system, including change control processes, accomplishes three major objectives:

- Establishes an evolutionary method to consistently identify and request changes to established baselines and to assess the value and effectiveness of those changes

- Provides opportunities to continuously validate and improve the project by considering the impact of each change

- Provides the mechanism for the project management team to consistently communicate all changes to the stakeholders

Integrated change control process includes some specific activities of the configuration management as summarized below:

- Configuration Identification: This provides the basis from which the configuration of products is defined and verified, products and documents are labeled, changes are managed, and accountability is maintained.

- Configuration Status Accounting: This involves capturing, storing, and accessing configuration information needed to manage products and product information effectively.

- Configuration Verification and Auditing: This involves confirming that the performance and functional requirements defined in the configuration documentation have been satisfied.

Under integrated change control, every documented requested change must be either accepted or rejected by some authority within the project management team or an external organization representing the initiator, sponsor, or customer. Integrated change control can, possibly, be controlled by a Change Control Board.

DEFINITION OF INPUTS TO STEP 6

The inputs to STEP 6 include the following items, which were all described earlier:

- Project management plan

- Requested changes

- Work performance information

- Recommended preventive actions

- Deliverables

DEFINITION OF TOOLS AND TECHNIQUES FOR STEP 6

- Project management methodology: This defines a process that helps a project management team in implementing integrated change control for the project.

- Project management information system: This is an automated system used by the team as an aid for the implementation of an integrated change control process for the project. It also facilitates feedback for the project and controls changes across the project.

- Expert judgment: This refers to the process whereby the project team uses stakeholders with expert judgment on the change control board to control and approve all requested changes to any aspect of the project.

DEFINITION OF OUTPUTS OF STEP 6

The outputs of STEP 6 include the following:

- Approved change requested

- Rejected change requests

- Project management plan (updates)

- Project scope statement (updates)

- Approved corrective actions

- Approved preventive actions

- Approved defect repair

- Validated defect repair

- Deliverables

TABLE 1.8   Tools and Techniques for Closing Project Within Integration
Management

STEP 7: Close Project

Inputs
Project management plan
Contract documentation
Enterprise environmental factors
Organizational process assets
Work performance information
Deliverables
Other in-house (custom) factors of relevance and interest

Tools and Techniques
Project management methodology
Project management information system
Expert judgment
Other in-house (custom) tools and techniques

Output(s)
Administrative closure procedure
Contract closure procedure
Final product, service or result
Updates on organizational process assets
Other in-house outputs, reports, and data inferences of interest to the
  organization

STEP 7: CLOSE PROJECT

At its completion, a project must be formally closed. This involves
performing the project closure portion of the project management
plan or closure of a phase of a multi-phase project. There are two
main procedures developed to establish interactions necessary to
perform the closure function:

- Administrative closure procedure: This provides details of
  all activities, interactions, and related roles and responsibilities involved in executing the administrative closure of
  the project. It also covers activities needed to collect project records, analyze project success or failure, gather lessons
  learned, and archive project information.

- Contract closure procedure: This involves both product verification and administrative closure for any existing contractual agreements. Contract closure procedure is an input to the close contract process.

DEFINITION OF INPUTS TO STEP 7

The inputs to STEP 7 are the following:

- Project management plan

- Contract documentation: This is an input used to perform the contract closure process and includes the contract itself as well as changes to the contract and other documentation, such as technical approach, product description, or deliverable acceptance criteria and procedures.

- Enterprise environmental factors

- Organizational process assets

- Work performance information

- Deliverables, as previously described, and also as approved by the integrated change control process.

DEFINITION OF TOOLS AND TECHNIQUES OF STEP 7

- Project management methodology
- Project management information system
- Expert judgment

DEFINITION OF OUTPUTS OF STEP 7

- Administrative closure procedure
  - Procedures to transfer the project products or services to production and/or operations are developed and established at this stage

- This stage covers a step-by-step methodology for administrative closure that addresses the following:

  - Actions and activities to define the stakeholder approval requirements for changes and all levels of deliverables

  - Actions and activities confirm project has met all sponsor, customer, and other stakeholders' requirements.

  - Actions and activities to verify the all deliverables have been provided and accepted.

  - Actions and activities to validate completion and exit criteria for the project.

- Contract Closure Procedure

  - This stage provides a step-by-step methodology that addresses the terms and conditions of the contracts and any required completion or exit criteria for contract closure.

  - Actions performed at this stage formally close all contracts associated with the completed project.

- Final Product, Service, or Result

  - Formal acceptance and handover of the final product, service, or result that the project was authorized to provide

  - Formal statement confirming that the terms of the contract have been met

- Organizational Process Assets (Updates)

  - Development of the index and location of project documentation using the configuration management system

- Formal acceptance documentation, which formally indicates the customer or sponsor has officially accepted the deliverables

- Project files, which contain all documentation resulting from the project activities.

- Project closure documents, which consist of a formal documentation indicating the completion of the project and transfer of deliverables

- Historical information, which is transferred to the knowledge base of lessons learned for use by future projects

- Traceability of process steps

PROJECT SUSTAINABILITY

Project efforts must be sustained in other for a project to achieve the intended end results in the long run. Project sustainability is not often addressed in project management; but it is essential, particularly for STE type of projects.

Sustainability, in ordinary usage, refers to the capacity to maintain a certain process or state indefinitely. In day-to-day parlance, the concept of sustainability is applied more specifically to living organisms and systems, particularly environmental systems. As applied to the human community, sustainability has been expressed as meeting the needs of the present without compromising the ability of future generations to meet their own needs. The term has its roots in ecology as the ability of an ecosystem to maintain ecological processes, functions, biodiversity, and productivity into the future. When applied to systems, sustainability brings out the conventional attributes of a system in terms of having the following capabilities:

- Self-regulation
- Self-adjustment

- Self-correction

- Self-recreation

To be sustainable, nature's resources must only be used at a rate at which they can be replenished naturally. Within the environmental science community, there is a strong belief that the world is progressing on an unsustainable path because the Earth's limited natural resources are being consumed more rapidly than they are being replaced by nature. Consequently, a collective human effort to keep human use of natural resources within the sustainable development aspect of the Earth's finite resource limits has become an issue of urgent importance. Unsustainable management of natural resources puts the Earth's future in jeopardy.

Sustainability has become a widespread, controversial, and complex issue that is applied in many different ways, including the following:

- Sustainability of ecological systems or biological organization (e.g., wetlands, prairies, forests)

- Sustainability of human organization (e.g., ecovillages, eco-municipalities, sustainable cities)

- Sustainability of human activities and disciplines (e.g., sustainable agriculture, sustainable architecture, sustainable energy)

- Sustainability of projects (e.g., operations, resource allocation, cost control)

For project integration, the concept of sustainability can be applied to facilitate collaboration across project entities. The process of achieving continuous improvement in operations, in a sustainable way, requires that engineers create new technologies that facilitate interdisciplinary thought exchanges. Under the STEP project methodology of this book, sustainability means asking questions that relate to the consistency and long-term execution of the project plan. Essential questions that should be addressed include the following:

- Is the project plan supportable under current operating conditions?

- Will the estimated cost remain stable within some tolerance bounds?

- Are human resources skills able to keep up with the ever-changing requirements of a complex project?

- Will the project team persevere toward the project goal, through both rough and smooth times?

- Will interest and enthusiasm for the project be sustained beyond the initial euphoria?

REFERENCES

PMI, *A Guide to the Project Management Body of Knowledge (PMBOK Guide)*, 6th Edition, Project Management Institute, Philadelphia, PA, 2017.

Badiru, A. B. (2014), "Quality Insights: The DEJI Model for Quality Design, Evaluation, Justification, and Integration," International Journal of Quality Engineering and Technology, Vol. 4, No. 4, pp. 369-378.

CHAPTER 2

Scope Management

SCOPE DEFINITIONS

It should be emphasized that project scope (PMI, 2017) differs from product scope (Badiru, 1993). Product scope describes the product to be delivered while project scope describes the work required to deliver the product. Product scope addresses the question of how a product works (how does the product work?); while project scope addresses the question of how a project performs (how did the project do?). Product scope is measured against product definition and specification. Project scope is measured against the project plan. Project scope is related to the project charter by the fact that the project charter specifies what is in the project scope. The budget and schedule of a project are directly impacted by the scope of the project.

SCOPE MANAGEMENT: STEP-BY-STEP IMPLEMENTATION

The five elements in the scope block diagram are carried out across the process groups. The overlay of the elements and the process groups are shown in Table 2.1. Thus, under the knowledge area of scope management, the required steps are:

Step 1: Scope Planning
Step 2: Scope Definition

Step 3: Create WBS

Step 4: Scope Verification

Step 5: Scope Control

Each step is carried out in a structure of Inputs – Tools and Techniques – Output analysis. Table 2.2 shows the input-to-output items for project scope planning. The tabular format is useful for explicitly identifying what the project analyst needs to do or use for each step of the project management process. Tables 2.3 through 2.6 present the input-to-output entries for the other steps under project scope management.

Project scope management plan provides guidance on how project scope will be defined, documented, verified, managed, and controlled. Components of the scope management plan include the following:

- Process to prepare a detailed project scope statement

- Process to create, maintain, and approve WBS

- Process for specifying formal verification and acceptance of deliverables

- Process to control change requests to detailed project scope statement

TABLE 2.1 Implementation of Project Scope Management Across Process Groups

Initiating
Planning
1. Scope Planning
2. Scope Definition
3. Create WBS
Executing
Monitoring and Controlling
1. Scope Verification
2. Scope Control
Closing

TABLE 2.2   Tools and Techniques for Project Scope Planning within Scope Management

STEP 1: Project Scope Planning

Inputs
Enterprise environmental factors
Organizational process assets
Project charter
Preliminary project scope statement
Project management plan
Other in-house (custom) factors of relevance and interest

Tools and Techniques
Expert judgment
Templates, forms, standards
Other in-house (custom) tools and techniques

Output(s)
Scope management plan
Other in-house outputs, reports, and data inferences of interest to the
organization

The primary purpose of project scope definition is to explore deeper details of stakeholder needs, wants, desires, and expectations for the purpose of creating project requirements. It requires an outline of additional constraints and assumptions for implementing the project. Product analysis develops a better understanding of the product of the project. Stakeholder analysis identifies the influence and interests of stakeholders and documents their needs in order to create project requirements. In addition, project scope definition provides a documented basis for making future project decisions and it covers the following:

- Justification: Why is the project needed?

- Product description: What is the expected product of the project?

- Boundaries: What is included and not included in the project?

- Constraints and Assumptions

- Deliverables: What are the project deliverables?

- Objectives: How will the success of the project be assessed?

Work breakdown structures (WBS) contain deliverables specified as nouns (not verbs), which are identifiable and tangible products of work. The work package in WBS shows the lowest level of WBS component that can be scheduled, cost estimated, monitored, and controlled. The descriptions of work components are compiled in the WBS dictionary. WBS subdivides major project deliverables or sub-deliverables into smaller, more manageable components.

TABLE 2.3   Tools and Techniques for Project Scope Definition within Scope Management

STEP 2: Project Scope Definition
Inputs
Organizational process assets
Project charter
Preliminary project scope statement
Project scope management plan
Approved change requests
Other in-house (custom) factors of relevance and interest
Tools and Techniques
Product analysis
Identification of alternatives
Expert judgment
Stakeholder analysis
SIPOC
DMAIC
Other in-house (custom) tools and techniques
Output(s)
Project scope statement
Requested change
Scope management plan (updated)
Other in-house outputs, reports, and data inferences of interest to the organization

TABLE 2.4   Tools and Techniques for Creating WBS within Scope Management

STEP 3: Create WBS
Inputs
Organizational process assets
Scope statement
Scope management plan
Approved change requests
Other in-house (custom) factors of relevance and interest
Tools and Techniques
WBS templates
Decomposition
Work partitioning
Other in-house (custom) tools and techniques
Output(s)
Scope statement (updated)
Work breakdown structure
WBS dictionary
Scope baseline
Scope management plan (updated)
Requested changes
Other in-house outputs, reports, and data inferences of interest to the organization

Project deliverables should be defined in a sufficient level of detail to facilitate the development of project activities with cost and duration estimates. If an item is not contained in the WBS, it is not going to be done. That is, it is outside the scope of the project. Every item in the WBS has a verifiable deliverable and it is assigned a unique identifier (code), has a statement of work, responsible organization, and a list of schedule milestones. The scope baseline for a project is composed of the approved detailed project scope statement, associated WBS, and WBS dictionary. Note that WBS is not a quality statement. It is a statement of what needs to be done; not how well it is done.

Scope verification consists of a formal acceptance of project scope. This is typically performed at the end of a phase or project.

TABLE 2.5   Tools and Techniques for Project Scope Verification within Scope Management

STEP 4: Project Scope Verification
Inputs
Scope statement
WBS dictionary
Scope management plan
Deliverables
Other in-house (custom) factors of relevance and interest
Tools and Techniques
Inspection
Other in-house (custom) tools and techniques
Output(s)
Accepted deliverables
Requested changes
Recommended corrective actions
Other in-house outputs, reports, and data inferences of interest to the organization

Stakeholder (i.e., sponsor, client, etc.) formally accepts the scope. The verification ties objectives to WBS. Formal acceptance of scope is documented even though acceptance may be conditional.

For scope control, we should identify changes using the WBS elements. We will evaluate the impact on cost, schedule, resources, and product quality. Stakeholder and project management authorize change. Successful scope management requires documentation. All aspects of the project must be documented in writing. Several industry-based tools and techniques can be used for the purpose of scope control.

USE OF DMAIC FOR SCOPE MANAGEMENT

Figure 2.1 shows an application of DMAIC (Define, Measure, Analyze, Improve, Control) concept to scope management process. DMAIC is a basic component of the Six Sigma methodology for improving work processes by eliminating defects and it complements the Lean Approach, which focuses on eliminating

TABLE 2.6   Tools and Techniques for Project Scope Control within Scope Management

STEP 5: Project Scope Control

Inputs
Scope statement
WBS
WBS dictionary
Scope management plan
Performance reports
Approved change requests
Work performance information
Other in-house (custom) factors of relevance and interest

Tools and Techniques
Change control system
Variance analysis
Re-planning
Configuration management system
Other in-house (custom) tools and techniques

Output(s)
Project scope statement (updates)
WBS updates
WBS dictionary updates
Scope baseline updates
Requested changes
Recommended corrective action
Organizational process assets
Project management plan updates
Other in-house outputs, reports, and data inferences of interest to the
  organization

waste in work processes. The Six Sigma methodology is widely used in industry and it is quickly finding a place in service and project enterprises. Six Sigma represents a set of practices that improve efficiency and eliminate defects in products and services. Applying DMAIC to project scoping can ensure that a project covers all the elements defined in the scope statement and only the elements defined in the scope statement.

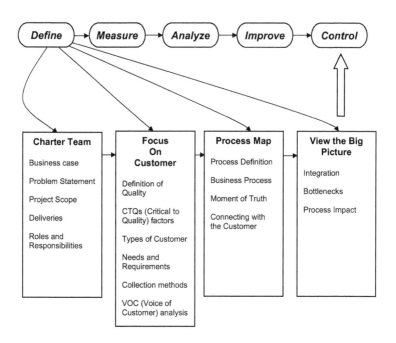

FIGURE 2.1   Application of DMAIC to process control.

The **Define** stage of DMAIC puts the project in the context of a specific business case. It is the first stage in the DMAIC process. In this stage, it is important to define specific goals in achieving outcomes that are consistent with both customer demands as well as the project organization's own business strategy. The define stage lays down a roadmap for project accomplishment. Definition may cover several items, but particular elements include team charter, focus on the customer, process map, and systems view of the big picture. Team charter covers business case, problem statement, project scope, deliveries, roles, and responsibilities. Also in the define stage, we focus on the customer or external constituents of the project. Items addressed in this respect include the definition of quality, CTQs (Critical To Quality) factors, types of customer, needs and requirements, data gathering methods, and VOC (Voice of Customer) analysis. The process map portion of Define maps out the steps and elements to be covered during a particular process. The items covered in this stage include process definition, business process outline, confronting the facts, and connecting with the customer. The major benefit of developing a process map is that it highlights the important tasks and functions to be undertaken during the process so that nothing is inadvertently ignored. Viewing the big picture during the Define stage of DMAIC implies using a systems view to analyze how each effort fits into the overall scheme of things.

The **Measure** stage of DMAIC lays the groundwork for measurement of the metrics of project performance. In order to determine whether or not defects have been reduced in a project's output, we need a base measurement. In this stage, accurate measurements must be made and relevant data must be collected and analyzed so that future comparisons can be measured to determine whether or not defects have been reduced. Procedures for measurement are particularly essential for control further down in a project. We must be able to measure an item before we can control or improve it.

The **Analyze** of DMAIC is very important to determine the relationships and the factors of causality in a project process. If the focus of a project is to generate products, services, or results, then we must understand what causes what and how the relationships can be enhanced.

The **Improve** stage of DMAIC outlines how to plan, pursue, and achieve improvement in the project process. Making improvements or optimizing processes inherent in a project, based on measurements and analysis, will ensure that defects are lowered and work processes are streamlined.

The **Control** stage is the last step in DMAIC methodology. Control ensures that any variances stand out and are corrected before they can adversely influence a process, thus, causing defects. Controls can be in the form of pilot runs to determine if the processes are capable and then once data is collected, a process can transition into the standard work process. Continued measurement and analysis must be undertaken to keep project work processes on track and free of defects below the Six Sigma quality limit.

USE OF SIPOC DIAGRAM FOR SCOPE MANAGEMENT

SIPOC (Suppliers, Inputs, Process, Outputs, Customers) is a diagram that is used to identify all project elements relevant for improvement before the project starts. The process improvement team may also add 'requirements' at the end of the SIPOC diagram to identify the specific customer requirements that are to be satisfied. This helps to obtain clarifications of what, who, what, why, and how of improvement efforts. Figure 2.2 shows a flow diagram for SIPOC.

The steps for building a SIPOC diagram are summarized below:

1. Select an area that is accessible or visible to the improvement team so that team members can post additions to the SIPOC diagram iteratively. This could be a presentation template projected onto a screen, flip charts with headings (S-I-P-O-C), or Post-it notes mounted onto a wall. Iteratively add items under each heading and proceed through four to five high-level iterations.

2. Identify the suppliers for the project.

3. Identify and annotate the inputs required from the suppliers.

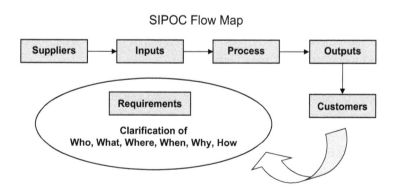

FIGURE 2.2   SIPOC flow diagram for process improvement.

4. Identify and map the process into which the inputs go. The inputs from the preceding step are required for the process to operate properly.

5. Identify and document the outputs of the process.

6. Identify the customers to whom the outputs are directed.

7. If desired, include the preliminary requirements of the customers for the purpose of clarifying who, what, when, where, why, and how aspects of the project. These clarifications will be verified during a later step of the Six Sigma measurement phase.

8. Discuss the SIPOC contents with the project sponsor, project champion, project management, and stakeholders of the project for verification and validation.

SIPOC helps to define a complex project to ensure that the project is in alignment with the scope statement. SIPOC is often applied at the Measure stage of the DMAIC methodology within the overall Six-Sigma effort. SIPOC complements and provides additional details for the usual process mapping and input-output scoping processes of project management. Table 2.7 shows examples of entries for a SIPOC Diagram example for technology dealership problem. Such a table can be expanded or customized for specific problems of interest. The SIPOC diagram is particularly useful for project scope verification by addressing the following questions:

- Who supplies inputs to the process?

- What specifications are placed on the inputs?

- Where will functions and operations be performed?

- Who are the true customers of the process?

- What are the requirements of the customers?

- When are requirements going to be satisfied?

- How will quality performance be ensured?

In a real-life project scoping environment, the scope can be quite volatile particularly for STEP projects where dynamic changes can be frequent and profound. Scopes are often enacted, repealed, re-enacted, modified, extended, altered, and adjusted for various reasons during the project lifecycle. This is where the structured approach of scope planning, scope definition, WBS creation, scope verification, and scope control is very useful for science, technology, and engineering projects.

SCOPE FEASIBILITY ANALYSIS

The scope feasibility of a project should be ascertained in terms of technical factors, managerial potential, economic factors, environmental issues, political expediency, financial analysis, economic realities, and community impact. Scope feasibility can be documented with a report showing all the ramifications of the project. This is particularly essential for multi-faceted science, technology, and engineering projects. Technical feasibility refers to the ability of the process to take advantage of the current state of the technology in pursuing further improvement. The technical capability of the personnel as well as the capability of the available technology should be considered. Managerial feasibility involves the capability of the management infrastructure of an organization to achieve and sustain the desired end-result. Management support, employee involvement, and commitment are key elements required to ascertain managerial feasibility.

Economic feasibility involves the ability of the proposed project to generate economic benefits. A benefit–cost analysis and a breakeven analysis are important aspects of evaluating the economic feasibility of a project from a scoping perspective. The tangible and intangible aspects of the project should be translated into economic terms to facilitate a consistent basis for assessment.

TABLE 2.7 Entries for SIPOC Diagram Example for Technology Vendor

Suppliers	Inputs	Process	Outputs	Customers	Requirements
Manufacturer	Deliveries	Client consultation	New client account	Technology users	Build to order
Suppliers	Option packages	Assessment of client	Purchase order	Technology showroom	Operating options
Repair service		needs	Paperwork to dealer	owners	Service contract
Outsource sites		Present options to	Paperwork to	Service departments	Installation
		clients	manufacturer		arrangement
		Demo	Payment process		
		Client agreement	Service contract		
			Service notifications		

Financial feasibility should be distinguished from economic feasibility. Financial feasibility involves the capability of the project organization to raise the appropriate funds needed to implement the proposed project and maintain it throughout its life cycle. Project financing can be a major obstacle in projects because of the level of capital required and the volatility of science and technology. Loan availability, credit worthiness, equity, and loan schedule are important aspects of financial feasibility analysis for project scoping purposes.

Cultural feasibility deals with the compatibility of the proposed project with the cultural atmosphere both from social culture, as well as the work environment. In working-class communities, functions must be integrated with the local cultural practices and beliefs. For example, an industry that requires the services of females must take into consideration the cultural norms affecting the position of women in the workplace in some countries. In rural areas, technology development efforts must not violate culturally sacred grounds that have religious or historical implications.

Social feasibility addresses the influences that a proposed project may have on the social system in the project environment. The ambient social structure may be such that certain categories of workers may be in short supply or nonexistent. The effect of the project on the social status of the project participants must be assessed to ensure compatibility. It should be recognized that workers in certain industries may have certain status symbols within the society.

Community feasibility refers to the general acceptance of the proposed project. Even in cases where a project is expected to lead to economic development of a community, there may still be discontent and opposition from local residents, particularly where the "eminent domain" doctrine is exercised by the government. Recent examples include the "bridge to no where" project in Alaska. Apparently, some aspects of the scope of the bridge project were not adequately vetted during the decision process. Another good example is the "Pipeline of Discontent" also called "Line of Conflict" in the Ohio Press. This involves the construction of a $5.6 billion, 1,679-mile Rock Express Pipeline (or REX) from Colorado's Rocky Mountain region through the Midwest on to the Eastern part of the USA. The pipeline is presented by the Federal Government as a crucial new component of America's energy infrastructure and a boon for consumers. But communities along the path of the pipeline vigorously oppose it. Contentious issues include eminent domain, safety, and environmental impact; in spite of the economic potential that the project would bring.

Safety feasibility is another important aspect that should be considered in project planning. Safety feasibility refers to an analysis of whether the project is capable of being implemented, operated, and sustained safely with minimal adverse effects on the environment and safety. Unfortunately, environmental impact assessment is often not adequately and deeply addressed in project development projects; often because of the pressure to get a project done before the technology changes.

A politically feasible project may be referred to as a "politically correct project." Political considerations often dictate the direction for a proposed project. This is particularly true for development projects that may have significant government inputs and political implications. For example, political necessity may be a source of support for a project regardless of the project's merits. On the other hand, worthy projects may face insurmountable opposition simply because of political factors. Political feasibility analysis requires an evaluation of the compatibility of project goals with the prevailing goals of the political system.

DIMENSIONS OF SCOPE FEASIBILITY

In general terms, the elements of a feasibility analysis for a project should cover the following items:

Need analysis: This indicates recognition of a need for the project. The need may affect the organization itself, another organization, the public, or the government. A preliminary study should be conducted to confirm and evaluate the need. A proposal of how the need may be satisfied is then developed. Pertinent questions that should be asked include:

Is the need significant enough to justify the proposed project?

Will the need still exist by the time the project is completed?

What are the alternate means of satisfying the need?

What is the economic impact of the need?

Process work: This is the preliminary analysis done to determine what will be required to satisfy the need. The work may be performed by a consultant who is a subject matter expert in the project field. The preliminary study often involves system models or prototypes. For STEP projects, artist's conception and scaled-down models may be used for illustrating the general characteristics of a process.

Engineering and design: This involves a detailed technical study of the proposed project. Written quotations are obtained from suppliers and subcontractors as needed. Technology capabilities are evaluated as needed. Product design, if needed, should be done at this stage.

Cost estimate: This involves estimating project cost to an acceptable level of accuracy. Levels of around minus five percent to plus fifteen percent are common at this level of a project plan. Both the initial and operating costs are included in the cost estimation. Estimates of capital investment, recurring, and nonrecurring costs should also be contained in the cost estimate document.

Financial analysis: This involves an analysis of the cash flow profile of the project. The analysis should consider recapitalization requirements, return on investment, inflation, sources of capital, payback periods, breakeven point, residual values, market volatility, and sensitivity. This is a critical analysis since it determines whether or not and when funds will be available to the project.

The project cash flow profile helps to support the economic and financial feasibility of the project.

Project impacts: This portion of scope feasibility analysis provides an assessment of the impact of the proposed project. Environmental, social, cultural, and economic impacts may be some of the factors that will determine how a project is perceived by stakeholders. The value-added potential of the project should also be assessed. A value-added tax may be assessed based on the price of a product and the cost of the raw material used in making the product. The tax collected may be viewed as a contribution to the government coffers for reinvestment in the science, technology, and engineering infrastructure of the nation.

Conclusions and recommendations: Scope feasibility analysis should end with the overall outcome of the project analysis. This may indicate an endorsement or disapproval of the project. If disapproved, potential remedies to make it right should be presented. Recommendations on what should be done should be included in this section of the scope feasibility report.

INDUSTRY CONVERSION FOR PROJECT SCOPING

The conversion of an existing industry to a new industry may be a possible approach to satisfy STEP development plans. Industries that are no longer meeting the needs of the society due to economic, social, cultural, or military (defense) requirements may have suitable alternate roles to play in new STEP project efforts. For example, economic and military reforms can be leveraged to develop opportunities for private industry in what, otherwise, would be military-oriented production facilities. Recent changes in the international security environment have prompted several nations to start to investigate how military technology may be converted for industrial purposes. New STEP projects should consider the possibility of industry conversion to achieve project goals. Recapitalization of old industry can be channeled toward the development of contemporary science, technology, and engineering industries.

ASSESSMENT OF LOCAL RESOURCES AND WORKFORCE

The feasibility of a STEP project should consider an assessment of the resources available locally to support the proposed production operations. Most nations and communities are blessed with abundant natural resources. The sad fact is that these resources are often underdeveloped and underutilized. When the resources are fully developed, it is often through exploitation by external organizations. For example, many countries that enjoyed oil discovery and a boom in the 1970s now face economic uncertainty due to problems in that sector of the economy. Lack of diversification to utilize local resources and local work force can spell doom when large-scale STEP projects are undertaken.

Blame is often placed on the lack of technical expertise when expressing the chagrin of failed STEP projects. But when the expertise is available, it is frequently underused, mis-applied, or misappropriated. It is true that local experts working individually accomplish nothing in the overwhelming bureaucracy that engulfs their expertise. In order for local experts to have an impact, there must be a coalition. A strong professional coalition is the only means of bringing about a meaningful change. Task forces should be set up to document the availability of local resources and their respective potentials to generate derivative products. The availability of skilled local work force should also be factored into project scope feasibility analysis. The products should be prioritized based on the pressing needs of the society and global market realities.

DEVELOPING SCOPE-BASED PROJECT PROPOSAL

Once a project is shown to be feasible along most of the appropriate dimensions of its operation, the next step is to issue a request for proposal (RFP) depending on the funding sources involved. Proposals are classified as either "solicited" or "unsolicited." Solicited proposals are those written in response to a request for a proposal while unsolicited ones are those written without a formal invitation from the funding source. Many companies prepare proposals in response to inquiries received from potential clients. Many proposals are written under competitive bids. If an RFP is issued, it should include statements about project scope, funding level, deliverables, performance criteria, and deadlines.

The purpose of the RFP is to identify companies that are qualified to successfully conduct the project in a cost-effective manner. But cost should not be the only basis for selecting the winning bid. Formal RFPs are sometimes issued to only a selected list of bidders who have been preliminarily evaluated as being qualified. These may be referred to as targeted RFPs. In some cases, general or open RFPs are issued and whoever is interested may bid for the project. This, however, has been found to be inefficient in some respect. Ambitious, but unqualified, organizations waste valuable time preparing losing proposals. The proposal recipient, on the other hand, spends much time reviewing and rejecting unqualified proposals. Open proposals do have proponents who praise their "equal opportunity" approach.

In practice, each organization has its own RFP format, content, and procedures. The request is called by different names including PI (procurement invitation), PR (procurement request), RFB (request for bid), or IFB (invitation for bids). In some countries, it is referred to as request for tender (RFT). Irrespective of the format used, an RFP should request information on bidder's costs, technical capability, management, and other characteristics. It should, in turn, furnish sufficient scope information on the expected work. A typical detailed RFP should include the following:

- Project background: Need, scope, preliminary studies, and results.

- Project deliverables and deadlines: Product, service, or results that are expected from the project, when the products are expected, and how the products will be delivered should be contained in this document.

- Project performance specifications: Sometimes, it may be more advisable to specify system requirements rather than rigid specifications. This gives the systems or project analysts the flexibility to utilize the most updated and cost-effective technology in meeting the requirements. If rigid specifications are given, what is specified is what will be provided regardless of cost and level of efficiency.

- Funding level: This is sometimes not specified because of nondisclosure policies or because of budget uncertainties. However, whenever possible, the funding level should be indicated in the requirements. This will help responders to map their input of resources to the expected payout for the effort.

- Reporting requirements: Project reviews, format, number and frequency of written reports, oral communication, financial disclosure, and other requirements should be specified.

- Contract administration: Guidelines for data management, proprietary work, intellectual property rights, progress monitoring, proposal evaluation procedure, requirements for inventions, trade secrets, copyrights, and so on should be included in the RFP.

- Special requirements (as applicable): Facility access restrictions, equal opportunity/affirmative actions, small business support, access facilities for the handicap, false statement penalties, cost sharing, compliance with government regulations, and so on should be included if applicable.

- Boilerplates (as applicable): These are special requirements that specify the specific ways certain project items are handled. Boilerplates are usually written based on organizational policy and are not normally subject to conditional changes. For example, an organization may have a policy that requires that no more than 50 percent of a contract award will be paid prior to the completion of the contract. Boilerplates are quite common in government-related projects. Thus, STEP projects may need boilerplates dealing with environmental impacts, social contribution, and financial requirements. These are issues that may not normally be considered within the science and technology hustle and bustle to get the job done.

PROPOSAL PREPARATION SCOPE

Whether responding to an RFP or preparing an unsolicited proposal, care must be taken to provide enough detail to permit an accurate assessment of a project proposal. The proposing organization will need to find out the following:

- Project time-frame

- Level of competition

- The agency's available budget

- The structure of the funding agency

- Point of contact (POC) within the agency

- Previous contracts awarded by the agency

- Exact procedures used in awarding contracts

- Nature of the work done by the funding agency

The project proposal should present a detailed plan for executing the proposed project. The proposal may be directed to a management team within the same organization or to an external organization. The proposal contents may be written in two parts: a Technical Section and a Management Section.

TECHNICAL SECTION OF PROJECT PROPOSAL

Project background

1. Organization's expertise in the project area

2. Project scope

3. Primary objectives

4. Secondary objectives

Technical approach

- Required technology

- Available technology

- Problems and their resolutions

- Work breakdown structure

Work statement

- Task definitions and list

- Expectations

Schedule

- Gantt charts

- Milestones

- Deadlines

Project deliverables
The value of the project

- Significance

- Benefit

- Impact

MANAGEMENT SECTION OF PROJECT PROPOSAL

Project staff and experience

- Personnel credentials

Organization

- Task assignment

- Project manager, liaison, assistants, consultants, etc.

Cost analysis

- Personnel cost

- Equipment and materials

- Computing cost

- Travel

- Documentation preparation

- Cost sharing

- Facilities cost

Delivery dates

- Specified deliverables

Quality control measures

- Rework policy

Progress and performance monitoring

- Productivity measurement

Cost-control measures

SCOPE BUDGET PLANNING

Scoping can be an expression of budgeting. Scope determines the extent of budgeting just as budgeting determines the extent of project scope. The budgeting approach employed for a project can be used to express the overall organizational policy and commitment. Budget often specifies the following:

- Performance measures

- Incentives for efficiency

- Project selection criteria

- Expressions of organizational policy

- Plans for how resources are to be expended

- Catalyst for productivity improvement

- Control basis for managers and administrators

- Standardization of operations within a given horizon

The preliminary effort in the preparation of a budget is the collection and proper organization of relevant data. The preparation of a budget for a project is more difficult than the preparation of budgets for regular and permanent organizational endeavors, however. While recurring endeavors usually generate historical data which serve as inputs to subsequent estimating functions, projects, on the other hand, are often one-time undertakings without the benefit of prior data. The input data for the budgeting process may include inflationary trends, cost of capital, standard cost guides, past records, and forecast projections. Budgeting may be done as top-down or bottom-up.

SCOPING TOP-DOWN

This involves collecting data from upper-level sources such as top and middle managers. The cost estimates supplied by the managers may come from their judgments, past experiences, or past data on similar project activities. The cost estimates are passed to lower-level managers, who then break the estimates down into specific work components within the project. These estimates may, in turn, be given to line managers, supervisors, and so on to continue the process. In the end, individual activity costs are developed. The top management presents the overall budget while the line worker generates specific activity budget requirements. One advantage of the top-down budgeting approach is that individual work elements need not be identified prior to approving the overall project budget. Another advantage of the approach is that the aggregate or overall project budget can be reasonably accurate even though specific activity costs may contain substantial errors.

SCOPING BOTTOM-UP

In bottom-up budgeting, elemental activities, their schedules, descriptions, and labor skill requirements are used to construct detailed budget requests. The line workers that are actually performing the activities are asked to supply cost estimates. Estimates are made for each activity in terms of labor time, materials, and machine time. The estimates are then converted to monetary values. The estimates are combined into composite budgets at each successive level up the budgeting hierarchy. If estimate discrepancies develop, they can be resolved through the intervention of senior management, junior management, functional managers, project managers, accountants, or financial consultants. Analytical tools such as learning-curve analysis, work sampling, and statistical estimation may be used in the budgeting process as appropriate to improve the quality of cost estimates. All component costs and departmental budgets are combined into an overall budget and sent to top management for approval. A common problem with bottom-up budgeting is that individuals tend to overstate their needs with the notion that top management may cut the budget by some percentage. It should be noted, however, that sending erroneous and misleading estimates will only lead to a loss of credibility.

ZERO-BASE SCOPING

This is another budgeting approach that bases the level of project funding on previous performance. It is normally applicable to recurring programs, especially those in the public sector. Accomplishments in past funding cycles are weighed against the level of resource expenditure. Programs that are stagnant in terms of their accomplishments relative to budget size do not receive additional budgets. Programs that have suffered decreasing yields are subjected to budget cuts or even elimination. By contrast, programs that have a record of accomplishments are rewarded with larger budgets. A major problem with zero-base budgeting is that it puts participants under tremendous pressure to perform data collection, organization, and program justification. So, much time may be spent documenting program accomplishments that productivity improvements on current projects may be compromised. Proponents of zero-base budgeting see it as a good approach to encourage managers and administrators to be more conscious of their management responsibilities. From a project control perspective, the zero-base budgeting approach may be useful in identifying and eliminating specific activities that have not contributed to project goals in the past.

PROJECT SCOPING WITH WBS

Work breakdown structure (WBS) refers to the itemization of a project for planning, scheduling, and control purposes. It essentially communicates the scope of the project by presenting the inherent components of a project in a structured block diagram or interrelationship flow chart. WBS shows the hierarchies of parts (phases, segments, milestone, etc.) of the project. The purpose of constructing a WBS is to analyze the elemental components of the project in detail. If a project is properly designed through the application of WBS at the project planning stage, it becomes easier to estimate cost and time requirements of the project. Project control is also enhanced by the ability to identify how components of the project link together within the scope of the project.

CRITERIA FOR PROJECT REVIEW

Some of the specific criteria that may be included in project review and selection are presented below:

- Cost reduction
- Customer satisfaction
- Process improvement
- Revenue growth
- Operational responsiveness
- Resource utilization
- Project duration
- Execution complexity
- Cross-functional efficiency
- Partnering potential

HIERARCHY OF SELECTION

In addition to evaluating an overall project, elements making up the project may need to be evaluated on the basis of the hierarchy presented below. This will facilitate achieving an integrated project management view of the organization's operations.

- System
- Program
- Task
- Work packages
- Activity

SIZING OF PROJECTS

Associating a size measure with an industrial project provides a means of determining the level of relevance and efforts required. A simple guideline is presented below:

- Major (over 60 man-months of effort)

- Intermediate (6 to 60 man-months)

- Minor (Less than 6 man-months)

PLANNING LEVELS

When selecting projects and its associated work packages, planning should be done in an integrative and hierarchical manner following the levels of planning presented below:

- Supra level (The big picture)

- Macro level (The operational picture)

- Micro level (The task-level picture)

REFERENCES

Badiru, A. B., *Managing Industrial Development Projects: A Project Management Approach*, Van Nostrand Reinhold, New York, 1993.

PMI, *A Guide to the Project Management Body of Knowledge (PMBOK Guide)*, 6th Edition, Project Management Institute, Philadelphia Square, PA, 2017.

CHAPTER 3

Time Management

T IME MANAGEMENT INVOLVES THE effective and efficient use of time to facilitate the execution of a project expeditiously. Time, in terms of project schedule, is often the most noticeable aspect of a project. Consequently, time management is of utmost importance in project management. This is even more critical for projects, which are subject to rapid changes in technology. The first step of good time management is to develop a project plan that represents the process and techniques needed to execute the project satisfactorily. The effectiveness of time management is reflected in schedule performance analysis. Hence, scheduling is a major focus in project management. Many people erroneously view schedule management as project management. But, in fact, schedule management is just one aspect of project management.

TIME MANAGEMENT: STEP-BY-STEP IMPLEMENTATION

The six elements in the time management block diagram are carried out across the process groups. The overlay of the elements and the process groups are shown in Table 3.1. Thus, under the knowledge area of time management, the required steps are:

Step 1: Activity Definition

Step 2: Activity Sequencing

Step 3: Activity Resource Estimating

Step 4: Activity Time Estimating

Step 5: Project Schedule Development

Step 6: Project Schedule Control

Project time management is defined as the set of processes required to accomplish timely completion of the project. As with other steps of project management, time management processes overlap and interact with other cross-functional processes within the knowledge areas. Time management is preceded by the development of a project management plan, which is an output of project integration management.

Each step of project time management is carried out in a structure of Inputs – Tools and Techniques – Output analysis. Table 3.2 shows the input-to-output items for activity definition. The tabular format is useful for explicitly identifying what the project analyst needs to do or use for each step of the project management process. Tables 3.3 through 3.7 present the input-to-output entries for the other steps under project time management.

TABLE 3.1   Implementation of Project Time Management Across Process Groups

	Initiating	Planning	Executing	Monitoring and Controlling	Closing
Project Time Management		1. Activity Definition 2. Activity Sequencing 3. Activity Resource Estimating 4. Activity Duration Estimating 5. Project Schedule Development		6. Project Schedule Control	

TABLE 3.2   Tools and Techniques for Activity Definition within Time
Management

STEP 1: Activity Definition

Inputs
Enterprise environmental factors
Organizational process assets
Project scope statement
WBS
WBS Dictionary
Project management plan
Other in-house (custom) factors of relevance and interest

Tools and Techniques
Project decomposition
Templates, forms, standards
Expert judgment
Planning component
Rolling wave planning
Other in-house (custom) tools and techniques

Output(s)
Activity list
Activity attributes
Milestone list
Requested changes
Other in-house outputs, reports, and data inferences of interest to the organization

Activity involves identifying the specific activities that need to be performed to produce the various project deliverables. Under tools and techniques, decomposition defines the final outputs as schedule activities versus deliverables found in the WBS structure. WBS elements are nouns that identify deliverables, while schedule activities are verbs indicating actions to be performed to accomplish work elements. Activity definition process identifies deliverables at the lowest level in the WBS. These are called work packages. Activity definition takes the work packages and subdivides or decomposes them into smaller components called schedule activities, which provide the basis for scheduling, executing, monitoring, and controlling during the project lifecycle.

TABLE 3.3   Tools and Techniques for Activity Sequencing within Time
Management

STEP 2: Activity Sequencing

Inputs
Project scope statement
Activity list
Activity attributes
Milestone list
Approved change requests
Other in-house (custom) factors of relevance and interest

Tools and Techniques
Precedence diagramming method (PDM)
Arrow diagramming method (ADM)
Schedule network templates
Dependency determination
Applying leads and lags
Process control charts
Critical chain
Other in-house (custom) tools and techniques

Output(s)
Network diagram
Activity list updates
Activity attributes updates
Requested changes
Other in-house outputs, reports, and data inferences of interest to the
  organization
Project schedule

Enterprise environmental factors include existing organizational culture, systems, database repository, infrastructure, standards, and organization structure. Organizational process assets
include standard processes, policies, guidelines, communication
requirements, financial controls, existing change controls, and risk
control. Rolling wave planning is a form of progressive elaboration
of work. In this case, near-term work is planned in detail while far-
term work in the future is planned for at a relatively high (or broad)
level. Milestone lists in the project network can be mandatory or

TABLE 3.4   Tools and Techniques for Activity Resource Estimating within Time Management

STEP 3: Activity Resource Estimating

Inputs
Enterprise environmental factors
Organizational process assets
Activity list
Activity attributes
Resource availability
Project management plan
Other in-house (custom) factors of relevance and interest

Tools and Techniques
Expert judgment
Analysis of alternatives
Project management software
Bottom-up estimating
Goal programming
Portfolio management
Balance scorecard
Other in-house (custom) tools and techniques

Output(s)
Activity resource requirements
Activity attributes updates
Resource breakdown structure
Resource calendar updates
Required changes
Other in-house outputs, reports, and data inferences of interest to the organization

optional. Activity sequencing involves identifying and documenting logical relationships among schedule activities. Logical sequencing should highlight the precedence relationship and appropriate leads and lags. The three basic types of precedence relationship are:

1. Technical precedence requirement

2. Procedural precedence requirement

3. Imposed precedence requirement

TABLE 3.5   Tools and Techniques for Activity Duration Estimating within
Time Management

STEP 4: Activity Duration Estimating

Inputs

Enterprise environmental factors

Organizational process assets

Project scope statement

Activity list

Activity attributes

Activity resource requirements

Resource calendar

Project management plan

Other in-house (custom) factors of relevance and interest

Tools and Techniques

Expert judgment

Analogous estimating

Parametric estimating

Three-point estimates

Reserve analysis

Process control charts

Goal programming

Other in-house (custom) tools and techniques

Output(s)

Activity duration estimates

Activity attribute updates

Other in-house outputs, reports, and data inferences of interest to the organization

Of the three types of precedence constraints, technical precedence is the most difficult to circumvent. The procedural precedence requirement can, in many cases, be relaxed due to prevailing work-flow flexibility. The imposed relationship is often due to resource-shortage impositions. Thus, if we can change our work-flow concepts and exercise resource allocation options, we may be able to achieve project schedule improvements.

A lead is the amount of time by which the start of an activity leads (or overlaps with) the activity's predecessor. Lag is the amount of time by which an activity waits (or lags behind) after the finish time of the activity's predecessor. Project scope statement includes product characteristics that can affect sequencing. Approved changes are authorized changes to the project schedule, budget, or scope. Considering time–cost–quality relationships, the scope axis is often represented by project performance, project quality, or project expectations.

CPM NETWORK SCHEDULING

Project scheduling is often the most visible step in the sequence of steps of project management. The two most common techniques of basic project scheduling are the critical path method (CPM) and program evaluation and review technique (PERT). The network of activities contained in a project provides the basis for scheduling the project and can be represented graphically to show both the contents and objectives of the project. Extensions to CPM and PERT include precedence diagramming method (PDM) and critical resource diagramming (CRD). These extensions were developed to take care of unique project scenarios and requirements. PDM technique permits the relaxation of strict precedence structures in a project so that the project duration can be compressed. CRD handles the project scheduling process by using activity-resource assignments as the primary focus for the scheduling process. This approach facilitates resource-based scheduling rather than activity-based scheduling so that resources can be more effectively assigned and utilized.

CPM network analysis procedures originated from the traditional Gantt Chart or bar chart developed during World War I. There have been several mathematical techniques for scheduling activities, especially where resource constraints are a major factor. Unfortunately, mathematical formulations are not generally practical due to the complexity involved in implementing them for realistically large projects. Even computer implementations of the complex mathematical techniques often become too cumbersome for real-time managerial decisions. Project network diagram is any schematic representation of the logical relationships among

TABLE 3.6   Tools and Techniques for Project Schedule Development within Time Management

---

STEP 5: Project Schedule Development

Inputs

Organizational process assets
Project scope statement
Activity list
Activity attributes
Project schedule network diagrams
Activity resource requirements
Resource calendar
Activity duration estimates
Project management plan risk register
Other in-house (custom) factors of relevance and interest

Tools and Techniques

Schedule network analysis
Critical path method (CPM)
Schedule compression
What-if scenario analysis
Resource leveling
Critical chain method
Project management software
Calendar coordination
Adjusting leads and lags
Schedule model
Critical chain
Process control charts
Other in-house (custom) tools and techniques

Output(s)

Project schedule
Schedule model data
Schedule baseline
Resource requirement updates
Activity attributes updates
Project calendar updates
Requested changes
Project management plan updates
Other in-house outputs, reports, and data inferences of interest to the
  organization

TABLE 3.7   Tools and Techniques for Project Schedule Control within Time Management

STEP 6: Project Schedule Control

Inputs
Schedule management plan
Schedule baseline
Performance reports
Approved change requests
Other in-house (custom) factors of relevance and interest

Tools and Techniques
Progress reporting
Schedule change control system
Performance measurement
Project management software
Variance analysis
Schedule comparison bar chart
Critical chain
Other in-house (custom) tools and techniques

Output(s)
Schedule model data updates
Schedule baseline updates
Performance measurements
Requested changes
Recommended corrective actions
Organizational process assets updates
Activity list updates
Activity attributes updates
Project management plan updates
Other in-house outputs, reports, and data inferences of interest to the organization

project schedule activities. The diagram is typically drawn from left to right. The two major types of network diagrams are:

- Arrow-Diagramming-Method (ADM) or Activity-On-Arrow (AOA)

- Precedence Diagramming Method (PDM) or Activity-On-Node (AON)

In the AOA approach, arrows are used to represent activities, while nodes represent starting and ending points of activities. In the AON approach, nodes represent activities while arrows represent precedence relationships. Time, cost, and resource requirement estimates are developed for each activity during the network planning phase and are usually based on historical records, time standards, forecasting, regression functions, or other quantitative models. In AOA networks, dummy activities are denoted by dashed arrows. Dummy activities have zero-time durations and zero resource requirements. Only start-to-finish dependency relationships are possible in AOA.

A basic CPM project network analysis is typically implemented in three phases:

- Network planning phase

- Network scheduling phase

- Network control phase

Network planning: In the network planning phase, the required activities and their precedence relationships are determined. Precedence requirements may be determined on the basis of the following:

- Physical constraints, which represent mandatory activity dependencies.

- Procedural requirements, which represent discretionary activity order or dependencies.

- Imposed limitations, which represent externally-imposed activity dependencies.

An example of a physical constraint is the requirement to erect walls before installing a roof. This is a technical limitation grounded in a fixed sequence and can hardly be overcome. Such constraints are inherent in the nature of the project and will be found in any project of the same type. Thus, there is a *hard* logic associated with physical activity dependencies.

An example of a procedural constraint is a project team preference to have morning meetings prior to starting work. This is defined based on a preferred logic of the team. It may be based on a proven process or best-practice process. This often creates an arbitrary float or slack times. Thus, there is a *soft* logic associated with discretionary activity dependencies.

An example of an external constraint is a relationship imposed between project-based and non-project-based activities, such as the requirement to obtain a building permit before starting construction work. Such dependencies are not within the control of the project team because they are externally imposed. Regulatory requirements, trade agreements, and contractual boilerplates are other sources of external dependencies. If we can remove regulatory impediments, we can accomplish relaxation of imposed precedence relationships.

Network scheduling is performed by using forward-pass and backward-pass computations. These computations give the earliest and latest starting and finishing times for each activity. The amount of "slack" or "float" associated with each activity is determined during these computations. The activity path that includes the least slack in the network is used to determine the critical activities. This path, being the longest path in the network, also determines the duration of the project. Resource allocation and time–cost trade-offs are sometimes performed during network scheduling.

Network control involves tracking the progress of a project on the basis of the network schedule and taking corrective actions when needed. An evaluation of actual performance versus expected performance determines deficiencies in the project progress. The advantages of project network analysis are presented below.

ADVANTAGES FOR COMMUNICATION

- Clarifies project objectives

- Establishes the specifications for project performance

- Provides a starting point for more detailed task analysis

- Presents a documentation of the project plan

- Serves as a visual communication tool

ADVANTAGES FOR CONTROL

- Presents a measure for evaluating project performance

- Helps determine what corrective actions are needed

- Gives a clear message of what is expected

- Encourages team interaction

ADVANTAGES FOR TEAM INTERACTION

- Offers a mechanism for a quick introduction to the project

- Specifies functional interfaces on the project

- Facilitates ease of task coordination

CRITICAL CHAIN ANALYSIS

Critical Chain is the *Theory of Constraints* (Goldratt, 1997; Woeppel, 2001) applied to project management specifically for managing and scheduling projects. Constraint management is based on the principle that the performance of a system's constraint will determine the performance of the entire system (Niven, 2002; Martin, 2007; PMI, 2017; Collins, 2001). If a project's characteristic constraint is effectively managed, the overall project will be effectively managed. This is analogous to the belief that the worst performer in an organization will dictate the performance of the organization. Similarly, the weakest link in a chain determines the strength of the chain. Because the overall operation is essentially a series of linkages of activities, one break in the linkage determines a break of the overall operation. That is, it takes only one negative to negate a series of positives: $(+)(+)(+)$ $(+)(-)(+)(+) = (-)$. Looking at this from a production point of view, the bottleneck operation determines the throughput of a production system. From a group operation point of view, the last passenger on a complimentary shuttle bus determines the departure time of the bus. What all these mean in the context of project scheduling is that focus should be on the critical activities in the project network diagram. This means that the critical chain is the most important focus. With respect to applying the theory of constraints, there are three types of constraints:

1. Paradigm constraint (policy-based)

2. Resource constraint (physical limitation)

3. Material constraint (imposition by project environment)

Each constraint type impacts the project differently. For project scheduling purposes, critical chain is used to generate several alterations to the traditional CPM/PERT network. All individual activity slacks (or "buffer") become the project buffer. Each team member, responsible for his or her component of the activity network, creates a duration estimate free from any padding. A typical approach is to estimate based on a 50% probability of success. All activities on the critical chain (path) and feeder chains (noncritical chains in the network) are then linked with minimal time padding. The project buffer now is aggregated, and some proportion of the saved time is added to the project. Even adding 50% of the saved time significantly reduces the overall project schedule while requiring team members to be concerned less with activity padding and more with task completion. Even if the project team members miss their delivery date 50% of the time, the overall effect on the project's duration is minimized because of the downstream aggregated buffer. Readers can refer to the Bibliography at the end of this chapter for further details on the application of Critical Chain.

The same approach can also be used for tasks that are not on the critical chain. Accordingly, all feeder path activities are reduced by the same order of magnitude and a feeder buffer is constructed for the overall non-critical chain of activities. It should be noted that critical chain distinguishes between its use of buffer and the traditional project network use of project slack. In CPM/PERT, project slack is a function of the overall completed activity network. In other words, slack is an outcome of the task dependencies, whereas critical chain buffer is used as an *a priori* (or advance) planning contingency that is based on a logical redesign of each activity and the application of an aggregated project buffer at the end of the project. The following deficiencies have been noted about critical chain vis-à-vis the tradition CPM/PERT network analysis:

1. Lack of project milestones makes coordinated scheduling, particularly with external suppliers, highly problematic. Critics point out that the lack of in-process project milestones adversely affects the ability to coordinate schedule dates with suppliers who provide the external delivery of critical components.

2. Although it may be true that critical chain brings increased discipline to project scheduling, efficient methods for applying this technique to a firm's portfolio of projects are unclear; that is, critical chain offers benefits on a project-by-project basis, but its usefulness at the overall integrated program level has not been ascertained. Furthermore, because critical chain requires dedicated resources in a multi-project environment where resources are shared, it is impossible to avoid multi-tasking, which adversely impacts its utility.

3. Evidence of its success is still almost exclusively anecdotal and based on single-case studies. There is no large-scale empirical research to verify its overall effectiveness.

In summary, because of the dynamism of technology and fast-paced scientific evolution, STEP projects particularly require new ways of analysis and scheduling activities. The buffering approach offered by critical chain analysis represents another way of looking at the problem. The next chapter deals with STEP project cost management.

REFERENCES

Collins, J. (2001), *Good to Great*, HarperCollins Publishers, New York, NY.

Goldratt, E. M. (1997), *Critical Chain*, The North River Press, Great Barrington, MA.

Martin, H. L. (2007), *Techonomics: The Theory of Industrial Evolution*, Taylor & Francis Group/CRC Press, Boca Raton, FL.

Niven, P. R. (2002), *Balanced Scorecard: Step-by-Step: Maximizing Performance and Maintaining Results*, John Wiley, New York, NY.

PMI. (2017), *A Guide to the Project Management Body of Knowledge (PMBOK Guide)*, 6th Edition, Project Management Institute, Philadelphia PA.

Woeppel, M. J. (2001), *Manufacturer's Guide to Implementing the Theory of Constraints*, St. Lucie Press, Boca Raton, FL.

CHAPTER 4

Cost Management

Follow not only the money, but also the technology also is a lesson that aptly typifies what project cost management epitomizes; as suggested by the quotes referenced at the beginning of this chapter. Cost management is a primary function in project management. Cost is a vital criterion for assessing project performance. Cost management involves having an effective control over project costs through the use of reliable techniques of estimation, forecasting, budgeting, and reporting. Cost estimation requires collecting relevant data needed to estimate elemental costs during the life cycle of a project. Cost planning involves developing an adequate budget for the planned work. Cost control involves a continual process of monitoring, collecting, analyzing, and reporting cost data. Martin, (2007) defines Techonomics as the study of how technology affects the economy and a theory of organizational evolution that results from technological advance fueled and selected by economic success. Project cost management is impacted by the state of technology and the concomitant cost factors. The primary components of cost management within any project undertaking are:

- Cost estimating

- Cost budgeting

- Cost control

Cost control must be exercised across the other elements of the project management knowledge areas. The technique of earned value management plays a major and direct role in cost management. The technique is covered in detail later in this chapter.

COST MANAGEMENT: STEP-BY-STEP IMPLEMENTATION

The three elements in the cost management block diagram are carried out across the process groups. The overlay of the elements and the process groups are shown in Table 4.1. Thus, under the knowledge area of cost management, the required steps are:

Step 1: Cost Estimation

Step 2: Cost Budgeting

Step 3: Cost Control

Tables 4.2 through 4.4 present the inputs, tools, techniques, and outputs of each step

TABLE 4.1   Implementation of Project Cost Management Across Process Groups

Initiating
Planning
1. Cost Estimating
2. Cost Engineering
Executing
Monitoring and Controlling
3. Cost Control
Closing

TABLE 4.2   Tools and Techniques for Cost Estimating within Project Cost Management

Inputs

Enterprise environmental factors

Organizational process assets

Project scope statement

WBS

WBS dictionary

Project management plan

Other in-house (custom) factors of relevance and interest

Tools and Techniques

Analogous estimating

Resource cost rates

Goal programming

Return on investment analysis

Bottom-up estimating

Parametric estimating

Project management cost software

Vendor bid analysis

Reserve analysis

Cost of quality

CMMI (Capability Maturity Model Integration)

Other in-house (custom) tools and techniques

Output(s)

Activity cost estimates

Activity cost supporting detail

Requested changes

Cost management plan (updates)

Other in-house outputs, reports, and data inferences of interest to the
   organization

PROJECT PORTFOLIO MANAGEMENT

Project portfolio management is the systematic application of the tools and techniques of management to the collection of the cost-based element of a project. Examples of project portfolios would be planned initiatives, ongoing projects, and ongoing support services, and investment in emerging technology. A formal project portfolio management strategy enables measurement and

TABLE 4.3   Tools and Techniques for Cost Budgeting within Project Cost Management

STEP 2: Cost Budgeting

Inputs

Project scope statement
Work breakdown structure
WBS dictionary
Activity cost estimates
Activity cost estimate supporting detail
Project schedule
Resource calendars
Contract
Cost management plan
Other in-house (custom) factors of relevance and interest

Tools and Techniques

Cost aggregation
Portfolio management
Reserve analysis
Parametric estimating
Funding limit reconciliation
Balance scorecard
Critical chain elements budgeting
Other in-house (custom) tools and techniques

Output(s)

Cost baseline
Project funding requirements
Cost management plan (updates)
Requested changes
Other in-house outputs, reports, and data inferences of interest to the organization

objective evaluation of investment scenarios. Some of the key aspects of an effective project portfolio management are:

1. Define the project, supporting program, and enabling system as well as the required portfolio.

2. Define business value and the desired return on investment and prioritize projects.

TABLE 4.4   Tools and Techniques for Cost Control within Project Cost
Management

STEP 3: Cost Control

Inputs
Cost baseline
Project funding requirements
Performance reports
Work performance information
Approved change requests
Project management plan
Other in-house (custom) factors of relevance and interest

Tools and Techniques
Process control charts
Cost change control system
Performance measurement analysis
Forecasting
Trend analysis
Project performance reviews
Project management software
Variance analysis
Variance management
Earned value management
Other in-house (custom) tools and techniques

Output(s)
Cost estimates (updates)
Cost baseline (estimates)
Performance measurements
Forecasted completion
Requested changes
Recommended corrective actions
Organizational process assets (updates)
Project management plan (updates)
Other in-house outputs, reports, and data inferences of interest to the
  organization

3. Define an overall project portfolio management methodology.

4. Delineate an overall project portfolio in translating strategy
   into results.

5. Introduce a balanced scorecard that synthesizes and integrates the numerous and complex metrics related to different portfolio management processes into one framework.

6. Clarify projects that will provide effective allocation and management of limited resources.

7. Introduce a progressive project assessment approach including initial project assessment, mid-cycle project assessment, and closing project assessment.

8. Employ quantitative techniques to objectively assess a project for its absolute merit and relative merit against other projects.

9. Utilize weighted scoring models to quantify intangible benefits of the project.

10. Evaluate project decision techniques that clarify choices involving both risks and opportunities.

11. Build a business case for each project and rank order projects based on strategic fit, risks, opportunities, and the changing nature of science and technology.

12. Establish criteria for phasing out a project when it is no longer serving the desired purpose.

PROJECT COST ELEMENTS

Cost management in a project environment refers to the functions required to maintain effective financial control of the project throughout its life cycle. There are several cost concepts that influence the economic aspects of managing industrial projects. Within a given scope of analysis, there will be a combination of different types of cost factors as defined below:

Actual Cost of Work Performed

The cost actually incurred and recorded in accomplishing the work performed within a given time period.

Applied Direct Cost

The amounts recognized in the time period associated with the consumption of labor, material, and other direct resources, without regard to the date of commitment or the date of payment. These amounts are to be charged to work-in-process (WIP) when resources are actually consumed, material resources are withdrawn from inventory for use, or material resources are received and scheduled for use within 60 days.

Budgeted Cost for Work Performed

The sum of the budgets for completed work plus the appropriate portion of the budgets for level of effort and apportioned effort. Apportioned effort is effort that by itself is not readily divisible into short-span work packages but is related in direct proportion to measured effort.

Budgeted Cost for Work Scheduled

The sum of budgets for all work packages and planning packages scheduled to be accomplished (including work-in-process), plus the amount of level of effort and apportioned effort scheduled to be accomplished within a given period of time.

Burdened Costs

Burdened costs are cost components that are fully loaded with overhead charges as well as other pertinent charges. This includes the cost of management and other costs associated with running the business.

Cost Baseline

The cost baseline is used to measure and monitor project cost and schedule performance. It presents a summation of costs by period. It is used to measure cost and schedule performance and sometimes called performance measurement baseline (PMB).

Diminishing Returns

The law of diminishing returns refers to the phenomenon of successively less output for each incremental resource input.

Direct Cost

Cost that is directly associated with actual operations of a project. Typical sources of direct costs are direct material costs and direct labor costs. Direct costs are those that can be reasonably measured and allocated to a specific component of a project.

Economies of Scale

This is a term referring to the reduction of the relative weight of the fixed cost in total cost, achieved by increasing the quantity of output. Economies of scale help to reduce the final unit cost of a product and are often simply referred to as the savings due to mass production.

Estimated Cost at Completion

This refers to the sum of actual direct costs, plus indirect costs that can be allocated to a contract, plus the estimate of costs (direct and indirect) for authorized work remaining to be done.

First Cost

The total initial investment required to initiate a project or the total initial cost of the equipment needed to start the project.

Fixed Cost

Costs incurred regardless of the level of operation of a project. Fixed costs do not vary in proportion to the quantity of output. Examples of costs that make up the fixed cost of a project are administrative expenses, certain types of taxes, insurance cost, depreciation cost, and debt servicing cost. These costs usually do not vary in proportion to the quantity of output.

Incremental Cost

The additional cost of changing the production output from one level to another. Incremental costs are normally variable costs.

Indirect Cost

This is a cost that is indirectly associated with project operations. Indirect costs are those that are difficult to assign to specific components of a project. An example of an indirect cost is the cost of computer hardware and software needed to manage project operations. Indirect costs are usually calculated as a percentage of a component of direct costs. For example, the direct costs in an organization may be computed as 10% of direct labor costs.

Life-Cycle Cost

This is the sum of all costs, recurring and nonrecurring, associated with a project during its entire life cycle.

Maintenance Cost

This is a cost that occurs intermittently or periodically for the purpose of keeping project equipment in good operating condition.

Marginal Cost

Marginal cost is the additional cost of increasing production output by one additional unit. The marginal cost is equal to the slope of the total cost curve or line at the current operating level.

Operating Cost

This is a recurring cost needed to keep a project in operation during its life cycle. Operating costs may consist of such items as labor, material, and energy costs.

Opportunity Cost

This refers to the cost of forgoing the opportunity to invest in a venture that, if it had been pursued, would have produced an economic advantage. Opportunity costs are usually incurred due to limited resources that make it impossible to take advantage of all investment opportunities. It is often defined as the cost of the best-rejected opportunity. Opportunity costs can also be incurred due to a missed opportunity rather than due to an intentional rejection. In many cases, opportunity costs are hidden or implied because they typically relate to future events that cannot be accurately predicted.

Overhead Cost

These are costs incurred for activities performed in support of the operations of a project. The activities that generate overhead costs support the project efforts rather than contributing directly to the project goal. The handling of overhead costs varies widely from company to company. Typical overhead items are electric power cost, insurance premiums, cost of security, and inventory carrying cost.

Standard Cost

This is a cost that represents the normal or expected cost of a unit of the output of an operation. Standard costs are established in advance. They are developed as a composite of several component costs, such as direct labor cost per unit, material cost per unit, and allowable overhead charge per unit.

Sunk Cost

Sunk cost is a cost that occurred in the past and cannot be recovered under the present analysis. Sunk costs should have no bearing on the prevailing economic analysis and project decisions. Ignoring sunk costs can be a difficult task for analysts. For example, if $950,000 was spent four years ago to buy a piece of equipment for a technology-based project, a decision on whether or not to replace the equipment now should not consider that initial cost. But uncompromising analysts might find it difficult to ignore that much money. Similarly, an individual making a decision on selling a personal automobile would typically try to relate the asking price to what was paid for the automobile when it was acquired. This is wrong under the strict concept of sunk costs.

Total Cost

This is the sum of all the variable and fixed costs associated with a project.

Variable Cost

This cost varies in direct proportion to the level of operation or quantity of output. For example, the costs of material and labor required to make an item will be classified as variable costs since they vary with changes in the level of output.

BASIC CASH-FLOW ANALYSIS

Economic analysis is performed when a choice must be made between mutually exclusive projects that compete for limited resources. The cost performance of each project will depend on the timing and levels of its expenditures. The techniques of computing cash-flow equivalence permit us to bring competing project cash flows to a common basis for comparison. The common basis depends on the prevailing interest rate. Two cash flows that are equivalent at a given interest rate will not be equivalent at a different interest rate. The basic techniques for converting cash flows from one point in time to another are presented in the following sections.

TIME VALUE OF MONEY CALCULATIONS

Cash-flow conversion involves the transfer of project funds from one point in time to another. The following notation is used for the variables involved in the conversion process:

i = interest rate per period

n = number of interest periods

P = a present sum of money

F = a future sum of money

A = a uniform end-of-period cash receipt or disbursement

G = a uniform arithmetic gradient increase in period-by-period payments or disbursements.

In many cases, the interest rate used in performing economic analysis is set equal to the minimum attractive rate of return (MARR) of the decision maker. The MARR is also sometimes referred to as *hurdle rate, required internal rate of return (IRR), return on investment (ROI),* or *discount rate.* The value of MARR is chosen for a project based on the objective of maximizing the economic performance of the project.

CALCULATIONS WITH COMPOUND AMOUNT FACTOR

The procedure for the single payment compound amount factor finds a future amount, F, that is equivalent to a present amount, P, at a specified interest rate, i, after n periods. This is calculated by the following formula:

$$F = P(1+i)^n$$

Example: A sum of $5,000 is deposited in a project account and left there to earn interest for 15 years. If the interest rate per year is 12%, the compound amount after 15 years can be calculated as follows:

$$F = \$5,000(1+0.12)^{15} = \$27,367.85.$$

CALCULATIONS WITH PRESENT VALUE FACTOR

Present value (PV or P), also called present worth, is the present-day at-hand value of a cash flow. The present value factor computes PV when F is given. The present value factor is obtained by solving for P in the equation for the compound amount factor. That is,

$$P = F(1+i)^{-n}$$

Supposing it is estimated that $15,000 would be needed to complete the implementation of a project five years from now, how much should be deposited in a special project fund now so that the fund would accrue to the required $15,000 exactly five years from now? If the special project fund pays interest at 9.2% per year, the required deposit would be:

$$P = \$15,000(1+0.092)^{-5} = \$9,660.03$$

Other complex cash-flow calculations are based on these above basic calculations. Examples of those other calculations are not included in this focus book. Many reference books are available on the topic.

PROJECT COST ESTIMATION

Cost estimation and budgeting help establish a strategy for allocating resources in project planning and control. Based on the desired level of accuracy, there are three major categories of cost estimation for budgeting: *order-of-magnitude estimates*, *preliminary cost estimates*, and *detailed cost estimates*. Order-of-magnitude cost estimates are usually gross estimates based on the experience and judgment of the estimator. They are sometimes called "ballpark" figures. These estimates are typically made without a formal evaluation of the details involved in the project. The level of accuracy associated with order-of-magnitude estimates can range from −50% to +50% of the actual cost. These estimates provide a quick way of getting cost information during the initial stages of a project. The estimation range is summarized as follows:

$$50\%(\text{Actual Cost}) \le \text{Order-of-Magnitude Estimate}$$

$$\le 150\%(\text{Actual Cost})$$

Preliminary cost estimates are also gross estimates, but with a higher level of accuracy. In developing preliminary cost estimates, more attention is paid to some selected details of the project. An example of a preliminary cost estimate is the estimation of expected labor cost. Preliminary estimates are useful for evaluating project alternatives before final commitments are made. The level of accuracy associated with preliminary estimates can range from −20% to +20% of the actual cost, as shown in the following section:

$$80\%\left(\text{Actual Cost}\right) \leq \text{Preliminary Estimate}$$

$$\leq 120\%\left(\text{Actual Cost}\right)$$

Detailed cost estimates are developed after careful consideration is given to all the major details of a project. Considerable time is typically needed to obtain detailed cost estimates. Because of the amount of time and effort needed to develop detailed cost estimates, the estimates are usually developed after a firm commitment has been made that the project will take off. Detailed cost estimates are important for evaluating actual cost performance during the project. The level of accuracy associated with detailed estimates normally ranges from −5% to +5% of the actual cost.

$$95\%\left(\text{Actual Cost}\right) \leq \text{Detailed Cost} \leq 105\%\left(\text{Actual Cost}\right)$$

There are two basic approaches to generating cost estimates. The first one is a variant approach, in which cost estimates are based on variations of previous cost records. The other approach is the generative cost estimation, in which cost estimates are developed from scratch without taking previous cost records into consideration.

OPTIMISTIC AND PESSIMISTIC COST ESTIMATES

Using an adaptation of the PERT formula, we can combine optimistic and pessimistic cost estimates. If O = optimistic cost estimate, M = most likely cost estimate, and P = pessimistic cost estimate, the estimated cost can be stated as follows:

$$E[C] = \frac{O + 4M + P}{6}$$

and the cost variance can be estimated as follows:

$$V[C] = \left[\frac{P - O}{6}\right]^2$$

PROJECT BUDGET ALLOCATION

Project budget allocation involves sharing limited resources among competing tasks in a project. The budget allocation process serves the following purposes:

- A plan for resource expenditure
- A project selection criterion
- A projection of project policy
- A basis for project control
- A performance measure
- A standardization of resource allocation
- An incentive for improvement

TOP-DOWN BUDGETING

Top-down budgeting involves collecting data from upper-level sources such as top and middle managers. The figures supplied by the managers may come from their personal judgment, past experience, or past data on similar project activities. The cost estimates are passed to lower-level managers, who then break the estimates down into specific work components within the project. These estimates may, in turn, be given to line managers, supervisors, and lead workers to continue the process until individual activity costs are obtained. Thus, top management provides the global budget, while the functional level worker provides specific budget requirements for project items.

BOTTOM-UP BUDGETING

In this method, elemental activities, and their schedules, descriptions, and labor skill requirements are used to construct detailed budget requests. Line workers familiar with specific activities are asked to provide cost estimates, and then make estimates for each activity in terms of labor time, materials, and machine time. The estimates are then converted to an appropriate cost basis. The dollar estimates are combined into composite budgets at each successive level up the budgeting hierarchy. If estimate discrepancies develop, they can be resolved through the intervention of senior management, middle management, functional managers, project manager, accountants, or standard cost consultants.

To further aid in the process, analytical tools such as learning-curve analysis, work sampling, and statistical estimation may be employed in the cost estimation and budgeting processes.

BUDGETING AND RISK ALLOCATION FOR TYPES OF CONTRACT

Budgeting and allocation of risk are handled based on the type of contract involved. The list below carries progressively higher risk to the buyer (customer) while it carries progressively lower risk to the contractor (producer):

Type 1: Firm fixed price (FFP)

Type 2: FFP with economic adjustment

Type 3: Fixed price incentive fee (FPIF)

Type 4: Cost and cost sharing (CCS)

Type 5: Cost plus incentive fee (CPIF)

Type 6: Cost plus award fee (CPFF)

Type 7: Cost plus fixed fee (CPFF)

Type 8: Cost plus percentage fee (CPPF)

Type 9: Indefinite delivery

Type 10: Time and materials

Type 11: Basic agreements (Blanket contract)

A Type 1 contract carries the highest risk to the contractor (producer) whereas it carries the lowest risk to the buyer (customer). A Type 11 contract carries the lowest risk to the contractor (producer) whereas it carries the highest risk to the buyer (customer). The risk level is progressive in each direction of the list.

COST MONITORING

As a project progresses, costs can be monitored and evaluated to identify areas of unacceptable cost performance. Graphical plots may be used to evaluate cost, schedule, and time performance of a project. An approach similar to the profit ratio presented earlier may be used along with the plot to evaluate the overall cost performance of a project over a specified planning horizon. Presented below is a formula for cost performance index (CPI):

$$CPI = \frac{\text{Area of cost benefit}}{\text{Area of cost benefit} + \text{area of cost overrun}}.$$

As in the case of the profit ratio, CPI may be used to evaluate the relative performances of several project alternatives or to evaluate the feasibility and acceptability of an individual alternative.

PROJECT BALANCE TECHNIQUE

One other approach to monitoring cost performance is the project balance technique. The technique helps in assessing the economic state of a project at a desired point in time in the life cycle of the project. It calculates the net cash flow of a project up to a given point in time. The project balance is calculated as follows:

$$B(i)_t = S_t - P(1+i)^t + \sum_{k=1}^{t} \text{PW}_{\text{income}}(i)_k$$

where:

$B(i)_t$ = project balance at time t at an interest rate of $i\%$ per period

PW income $(i)_k$ = present worth of net income from the project up to time k

P = initial cost of the project

S_t = salvage value at time t.

The project balance at time t gives the net loss or net profit associated with the project up to that time. Some of the factors influencing schedule, performance, and cost problems are summarized below; with suggested lists of control actions:

CAUSES OF SCHEDULE PROBLEMS

- Delay of critical activities
- Unreliable time estimates
- Technical problems
- Precedence structure
- Change of due dates
- Bad time estimates
- Changes in management direction

SCHEDULE CONTROL ACTIONS

- Use activity crashing
- Redesign tasks
- Revise milestones
- Update time estimates
- Change the scope of work
- Combine related activities
- Eliminate unnecessary activities (i.e., operate lean)

CAUSES OF PERFORMANCE PROBLEMS

- Poor quality
- Poor functionality
- Maintenance problems
- Poor mobility (knowledge transfer)
- Lack of training
- Lack of clear objectives

PERFORMANCE CONTROL ACTIONS

- Use SMART job objectives (Specific, Measurable, Aligned, Realistic, Timed)
- Use improved tools/technology
- Adjust project specifications
- Improve management oversight
- Review project priorities
- Modify project scope

- Allocate more resources
- Require higher level of accountability
- Improve work ethics (Through training, mentoring, and education)

CAUSES OF COST PROBLEMS

- Inadequate budget
- Effects of inflation
- Poor cost reporting
- Increase in scope of work
- High overhead cost
- High labor cost

COST CONTROL ACTIONS

- Reduce labor costs
- Use competitive bidding
- Modify work process
- Adjust work breakdown structure
- Improve coordination of project functions
- Improve cost estimation procedures
- Use less expensive raw materials
- Mitigate effects of inflationary trends (e.g., use of price hedging in procurement)
- Cut overhead costs
- Outsource work

ELEMENTS OF COST CONTROL

Cost control, in the context of cost management, refers to the process of regulating or rectifying cost attributes to bring them within acceptable levels (Collins, 2001; PMI, 2017; Dettmer, 1997; Goldratt, 1997; Niven, 2002; Woeppel, 2001). Because of the volatility and dynamism often encountered in complex projects, it is imperative to embrace the following project cost control practices as presented in PMBOK:

- Influence the factors that create changes to the cost baseline

- Ensure requested changes are agreed upon

- Manage the actual changes when and as they occur

- Assure that potential cost overruns do not exceed authorized funding (by period and in total)

- Monitor cost performance to detect and understand variances from the cost baseline

- Record all appropriate changes accurately against the cost baseline

- Prevent incorrect, inappropriate, or unapproved changes from being included in cost reports

- Inform appropriate stakeholders or approved changes

- Act to bring expected cost overruns within acceptable limits

- Use the earned value technique (EVT) to track and rectify cost performance.

CONTEMPORARY EARNED VALUE TECHNIQUE

This section details the elements of a contemporary earned value technique (EVT). EVT is used primarily for cost control purposes. The technique involves developing important diagnostic values for each schedule activity, work package, or control element. Although the definitions presented below are similar to those in the forgoing C/SCSC discussions, there are shades of differences that are important to highlight. The definitions according to PMI's PMBOK are summarized below:

Planned value (PV): This is the budgeted cost for the work scheduled to be completed on an activity or WBS element up to a given point in time.

Earned value (EV): This is the budgeted amount for the work actually completed on the schedule activity or WBS component during a given time period.

Actual cost (AC): This is the total cost incurred in accomplishing work on the schedule activity or WBS component during a given time period. AC must correspond in definition, scale, units, and coverage to whatever was budgeted for PV and EV. For example, direct hours only, direct costs only, or all costs including indirect costs.

The PV, EV, and AC values are used jointly to provide performance measures of whether or not work is being accomplished as planned at any given point in time. The common measures of project assessment are cost variance (CV) and schedule variance (SV).

Cost variance (CV): This equals earned value minus actual cost. The cost variance at the end of the project will be the difference between the budget at completion (BAC) and the actual amount expended.

$$CV = EV - AC$$

Schedule variance (SV): This equals earned value minus planned value. Schedule variance will eventually become zero when the project is completed because all of the planned values will have been earned.

$$SV = EV - PV$$

Cost performance index (CPI): This is an efficiency indicator relating earned value to actual cost. It is the most commonly used cost-efficiency indicator. CPI value less than 1.0 indicates a cost overrun of the estimates. CPI value greater than 1.0 indicates a cost advantage (under-run) of the estimates.

$$CPI = \frac{EV}{AC}$$

Cumulative CPI (CPI^C): This is a measure that is widely used to forecast project costs at completion. It equals the sum of the periodic earned values (Cum. EV) divided by the sum of the individual actual costs (Cum. AC).

$$CPI^C = \frac{EV^C}{AC^C}$$

Schedule performance index (SPI): This is a measure that is used to predict the completion date of a project. It is used in conjunction with CPI to forecast project completion estimates.

$$SPI = \frac{EV}{PV}$$

Estimate to complete (ETC) based on new estimate: Estimate to complete equals the revised estimate for the work remaining as determined by the performing organization. This is an independent non-calculated estimate to complete for all the work remaining. It considers the performance or production of the resources to date. The calculation of ETC uses two alternate formulas based on earned value data.

ETC based on atypical variances: This calculation approach is used when current variances are seen as *atypical* and the expectations of the project team are that similar variances will *not* occur in the future.

$$ETC = BAC - EV^C$$

Where BAC = Budget at completion

ETC based on typical variances: This calculation approach is used when current variances are seen as *typical* of what to expect in the future.

$$ETC = \frac{BAC - EV^{C}}{CPI^{C}}$$

Estimate at completion (EAC): This is a forecast of the most likely total value based on project performance. EAC is the projected or anticipated total final value for a schedule activity, WBS component, or project when the defined work of the project is completed. One EAC forecasting technique is based upon the performance organization providing an estimate at completion. Two other techniques are based on earned value data. The three calculation techniques are presented below. Each of the three approaches can be effective for any given project because it can provide valuable information and signal if the EAC forecasts are not within acceptable limits.

EAC using a new estimate: The approach calculates the actual costs to date plus a new ETC that is provided by the performing organization. This is most often used when past performance shows that the original estimating assumptions were fundamentally flawed or that they are no longer relevant due to a change in project operating conditions.

$$EAC = AC^{C} + ETC$$

EAC using remaining budget: In this approach, EAC is calculated as cumulative actual cost plus the budget that is required to complete the remaining work; where the remaining work is the budget at completion minus the earned value. This approach is most often used when current variances are seen as *atypical* and the project management team expectations are that similar variances will not occur in the future.

$$EAC = AC^C + (BAC - EV)$$

Where (BAC – EV) = remaining project work = remaining PV.

EAC using cumulative CPI: In this approach, EAC is calculated as actual costs to date plus the budget that is required to complete the remaining project work, modified by a performance factor. The performance factor of choice is usually the cumulative CPI. This approach is most often used when current variances are seen as *typical* of what to expect in the future.

$$EAC = AC^C + \frac{(BAC - EV)}{CPI^C}$$

Other important definitions and computational relationships among the earned value variables are:

Earned → Budgeted cost of work actually performed

Planned → Budgeted cost of work scheduled

Actual → Cost of actual work performed

Ending CV = Budget at completion – Actual amount spent at the end

          = BAC – EAC

          = VAC (Variance at Completion)

EAC   = ETC + AC

          = (BAC – EV) + AC

          = AC + (BAC – EV)

ETC   = EAC – AC

          = BAC - EV

Activity-based Costing

Activity-based costing (ABC) has emerged as an effective costing technique for industrial projects. The major motivation for ABC is that it offers an improved method to achieve enhancements in operational and strategic decisions. Activity-based costing offers a mechanism to allocate costs in direct proportion to the activities that are actually performed. This is an improvement over the traditional way of generically allocating costs to departments. It also improves the conventional approaches to allocating overhead costs. In general, activity-based costing (ABC) is a method for estimating the resources required to operate an organization's business activities, produce its products, and provide services to its clients.

The ABC methodology assigns resource costs through activities to the products and services provided to its customers. It is generally used as a tool for understanding product and customer costs with respect to project profitability. ABC is also frequently used to formulate strategic decisions such as product pricing, outsourcing, and process improvement efforts.

The use of PERT/CPM, precedence diagramming, the critical resource diagramming method, and work breakdown structure (WBS) can facilitate the decomposition or breakdown of a task to provide information for activity-based costing. Some of the potential impacts of ABC on a production line include the following:

- Identification and removal of unnecessary costs
- Identification of the cost impact of adding specific attributes to a product
- Indication of the incremental cost of improved quality
- Identification of the value-added points in a production process
- Inclusion of specific inventory carrying costs
- Provision of a basis for comparing production alternatives
- Ability to assess "what-if" scenarios for specific tasks.

ABC is just one component of the overall activity-based management in an organization, and thus has its limitations as well. Activity-based management involves a more global management approach to the planning and control of organizational endeavors. This requires consideration for product planning, resource allocation, productivity management, quality control, training, line balancing, value analysis, and a host of other organizational responsibilities. In the implementation of ABC, several issues must be considered:

- Level and availability of resources committed to developing activity-based information and cost

- Duration and level of effort needed to achieve ABC objectives

- Level of cost accuracy that can be achieved by ABC

- Ability to track activities based on ABC requirements

- Challenge of handling the volume of detailed information provided by ABC

- Sensitivity of the ABC system to changes in activity configuration

From activity-based management (ABM) to activity-based costing (ABC), there are both qualitative as well as quantitative aspects of tracking, managing, and controlling costs. Unfortunately, many attempts to use ABC often degenerate into conceptual arm-waving rather than real quantitative accountability. To be successful, the same SMART principle that was discussed previously can be applied for developing ABC strategies. Under ABM and ABC, cost tracking must satisfy the following SMART requirements:

Specific: Cost tracking must be specific so as to facilitate accountability.

Measurable: Cost tracking must be measurable.

Aligned: Cost tracking must be aligned with the organization's goals.

Realistic: Cost tracking must be realistic and within the organization's capability.

Timed: Cost tracking must be timed in order to avoid ambiguities.

Also, to increase the effectiveness of ABC, an organization should use parametric cost techniques, which utilize project characteristics (parameters) to develop mathematical models for cost management. In summary, project cost management requires more prudent approaches compared to conventional cost management practices. Frequent changes in science, technology, and engineering undertakings lead to the dynamism of cost scenarios. Consequently, step-by-step tractable approaches must be used.

REFERENCES

Collins, J. (2001), *Good to Great*, Harper Collins Publishers, New York, NY.

Dettmer, H. W. (1997), *Goldratt's Theory of Constraints: A Systems Approach to Continuous Improvement*, Quality Press, Milwaukee, WI.

Goldratt, E. M. (1997), *Critical Chain*, The North River Press, Great Barrington, MA.

Martin, H. L. (2007), *Techonomics: The Theory of Industrial Evolution*, Taylor & Francis Group/CRC Press, Boca Raton, FL.

Niven, P. R. (2002), *Balanced Scorecard: Step-by-Step: Maximizing Performance and Maintaining Results*, John Wiley, New York, NY.

PMI (2004), *A Guide to the Project Management Body of Knowledge (PMBOK Guide)*, 3rd Edition, Project Management Institute, Newtown Square, PA.

Woeppel, M. J. (2001), *Manufacturer's Guide to Implementing the Theory of Constraints*, St. Lucie Press, Boca Raton, FL.

CHAPTER 5

Quality Management

PROJECT QUALITY MANAGEMENT IS the next stage of the structural approach to project management in PMBOK guidelines (Badiru and Ayeni, 1993; PMI, 2017). Quality management involves ensuring that the performance of a project conforms to specifications with respect to the requirements and expectations of the project stakeholders and participants. The objective of quality management is to minimize deviation from the actual project plans. Quality management must be performed throughout the life cycle of a project not just by a final inspection of the product.

QUALITY MANAGEMENT: STEP-BY-STEP IMPLEMENTATION

The quality management component of the project management body of knowledge consists of the elements shown below. The three elements are carried out across the process groups presented earlier in this book. The overlay of the elements and the process groups are shown in Table 5.1. Thus, under the knowledge area of quality management, the required steps are:

Step 1: Perform Quality Planning

Step 2: Perform Quality Assurance

Step 3: Perform Quality Control

Tables 5.2 through 5.4 present the inputs, tools, techniques, and outputs of each step.

TABLE 5.1   Implementation of Project Quality Management Across Process Groups

Initiating
Planning
   1. Perform Quality Planning
Executing
   1. Perform Quality Assurance
Monitoring and Controlling
   1. Perform Quality Control
Closing

TABLE 5.2   Tools and Techniques for Quality Planning within Project Quality Management

STEP 1: Perform Quality Planning

Inputs
Enterprise environmental factors
Organizational process assets
Project scope statement
Project management plan
Other in-house (custom) factors of relevance and interest

Tools and Techniques
Cost/Benefit analysis
Benchmarking
Design of experiments
Cost of Quality (COQ) assessment
Group decision techniques
Other in-house (custom) tools and techniques

Output(s)
Quality management plan
Quality metrics
Quality check lists
Process improvement plan
Quality baseline
Project management plan (updates)
Other in-house outputs, reports, and data inferences of interest to the
   organization

TABLE 5.3   Tools and Techniques for Quality Assurance within Project
Quality Management

STEP 2: Perform Quality Assurance

Inputs
Quality management plan
Quality metrics
Process improvement plan
Work performance information
Approved change requests
Quality control measurements
Implemented change requests
Implemented corrective actions
Implemented defect repair
Implemented preventive repair
Other in-house (custom) factors of relevance and interest

Tools and Techniques
Quality planning tools and techniques
Quality audits
Process analysis
Quality control tools and techniques
Other in-house (custom) tools and techniques

Output(s)
Requested changes
Recommended corrective actions
Organizational process assets (updates)
Project management plan (updates)
Other in-house outputs, reports, and data inferences of interest to the
  organization

Improvement programs have the propensity to drift into anecdotal, qualitative, and subjective processes. Having a quantifiable and measurable approach helps to overcome this deficiency.

SIX SIGMA AND QUALITY MANAGEMENT

The Six Sigma approach, which was originally introduced by Motorola's Government Electronics Group, has caught on quickly in industry. Many major companies now embrace the approach as

TABLE 5.4   Tools and Techniques for Quality Control within Project Quality Management

STEP 3: Perform Quality Control

Inputs

Quality management plan

Quality metrics

Quality check lists

Organizational process assets

Work performance information

Approved change requests

Deliverables

Other in-house (custom) factors of relevance and interest

Tools and Techniques

Cause and effect diagram

Control charts

Flowcharting

Histogram

Pareto chart

Run chart

Scatter diagram

Statistical sampling

Quality inspection

Defect repair review

Other in-house (custom) tools and techniques

Output(s)

Quality control measurements

Validated defect repair

Quality baseline (updates)

Recommended corrective actions

Recommended preventive actions

Requested changes

Recommended defect repair

Organization process assets (updates)

Validated deliverables

Other in-house outputs, reports, and data inferences of interest to the organization

TABLE 5.5   Interpretation of ± Sigma Intervals from Mean

Process Quality Range	Percentage Coverage	Interpretation of Standard
1 Sigma	68.26%	Poor performance
2 Sigma	95.46%	Below expectation
3 Sigma	99.73%	Historical acceptable standard
4 Sigma	99.9937%	Contemporary
6 Sigma	99.99999985%	New competitive standard

the key to high-quality industrial productivity. Six sigma means six standard deviations from a statistical performance average. The Six Sigma approach allows for no more than 3.4 defects per million parts in manufactured goods or 3.4 mistakes per million activities in a service operation. To appreciate the effect of the Six Sigma approach, consider a process that is 99% perfect. That process will produce 10,000 defects per million parts. With Six Sigma, the process will need to be 99.99966% perfect in order to produce only 3.4 defects per million. Thus, Six Sigma is an approach that pushes the limit of perfection. Table 5.5 summarizes sigma ranges and process percentage coverage levels.

TAGUCHI LOSS FUNCTION

The philosophy of Taguchi loss function defines the concept of how deviation from an intended target creates a loss in the production process. Taguchi's idea of product quality analytically models the loss to the society from the time a product is shipped to customers. Taguchi loss function measures this conjectured loss with a quadratic function known as Quality Loss Function (QLF), which is mathematically represented as:

$$L(y) = k(y - m)^2$$

Where k is a proportionality constant, m is the target value, and y is the observed value of the quality characteristic of the product in question. The quantity, $(y - m)$, represents the deviation from the target. The larger the deviation, the larger the loss to the society. The constant k can be determined if $L(y)$, y, and m are known. Loss, in the QLF concept, can be defined to consist of several components. Examples of loss are provided below:

- **Opportunity cost** of not having the service of the product due to its quality deficiency. The loss of service implies that something that should have been done to serve the society could not be done.

- **Time lost** in the search to find (or troubleshoot) the quality problem.

- **Time lost** (after finding the problem) in the attempt to solve the quality problem. The problem identification effort takes away some of the time that could have been productively used to serve the society. Thus, the society incurs a loss.

- **Productivity loss** that is incurred due to the reduced effectiveness of the product. The decreased productivity deprives the society of a certain level of service and, thereby, constitutes a loss.

- **Actual cost** of correcting the quality problem. This is, perhaps, the only direct loss that is easily recognized. But there are other subtle losses that the Taguchi method can help identify.

- **Actual loss** (e.g., loss of life) due to a failure of the product resulting from its low quality. For example, a defective automobile tire creates a potential for traffic fatality.

- **Waste** that is generated as a result of lost time and materials due to rework and other non-productive activities associated with low quality of work.

IDENTIFICATION AND ELIMINATION OF SOURCES OF DEFECTS

The approach uses statistical methods to find problems that cause defects. For example, the total yield (number of nondefective units) from a process is determined by a combination of the performance levels of all the steps making up the process. If a process consists of 20 steps and each step is 98 percent perfect, then the performance of the overall process will be:

$$(0.98)^{20} = 0.667608 \, (\text{i.e., } 66.7608\%)$$

Thus, the process will produce 332,392 defects per million parts. If each step of the process is pushed to the Six Sigma limit, then the process performance will be:

$$(0.9999966)^{20} = 0.999932 \, (\text{i.e., } 99.9932\%)$$

Thus, the Six Sigma process will produce only 68 defects per million parts. This is a significant improvement over the original process performance. In many cases, it is not realistic to expect to achieve the Six Sigma level of production. But the approach helps to set a quality standard and provides a mechanism for striving to reach the goal. In effect, the Six Sigma process means changing the way workers perform their tasks so as to minimize the potential for defects.

The success of Six Sigma in industry ultimately depends on industry's ability to initiate and execute Six Sigma projects effectively. Thus, the project management approaches presented in this book are essential for realizing the benefits of Six Sigma. Project planning, organizing, team building, resource allocation, employee training, optimal scheduling, superior leadership, shared vision, and project control are all complementarily essential to implementing Six Sigma successfully. These success factors are not mutually exclusive. In many organizations, far too much focus is directed toward the statistical training for Six Sigma at the expense of proper project management development. This explains why many organizations have not been able to achieve the much-touted benefits of Six Sigma.

The success of the Toyota production system is not due to any special properties of the approach, but rather due to the consistency, persistence, and dedication of Toyota organizations in building their projects around all the essential success factors. Toyota focuses on changing the organizational mindset that is required in initiating and coordinating the success factors throughout the organization. Six Sigma requires the management of multiple projects with an identical mindset throughout the organization. The success of this requirement is dependent on the proper application of project management tools and techniques.

ROLES AND RESPONSIBILITIES FOR SIX SIGMA

Human roles and responsibilities are crucial in executing Six Sigma projects. The different categories of team players are explained below:

Executive Leadership: Develops and promulgates vision, direction. Leads change and maintain accountability for organizational results (on a full-time basis).

Employee Group: Includes all employees, supports organizational vision, receives and implements Six Sigma specs, serves as points of Total Process Improvement (TPM), exports mission statement to functional tasks, and deploys improvement practices (on full-time basis).

Six Sigma Champion: Advocates improvement projects, leads business direction, and coordinates improvement projects (on a full-time basis).

Six Sigma Project Sponsor: Develops requirements, engages project teams, leads project scoping, and identifies resource requirements (on part-time basis).

Master Belt: Trains and coaches Black Belts and Green Belts, leads large projects, and provides leadership (on full-time basis).

Black Belt: Leads specific projects, facilitates troubleshooting, coordinates improvement groups, and trains and coaches project team members (on full-time basis).

Green Belt: Participates on Black Belt teams, leads small projects (on part-time project-specific basis).

Six Sigma Project Team Members: Provides specific operational support, facilitates inward knowledge transfer, and links to functional areas (on part-time basis).

Statistical Techniques for Six Sigma

Statistical process control (SPC) means controlling a process statistically. SPC originated from the efforts of the early quality control researchers. The techniques of SPC are based on basic statistical concepts normally used for statistical quality control. In a manufacturing environment, it is known that not all products are made exactly alike. There are always some inherent variations in units of the same product. The variation in the characteristics of a product provides the basis for using SPC for quality improvement. With the help of statistical approaches, individual items can be studied and general inferences can be drawn about the process or batches of products from the process. Since 100% inspection is difficult or impractical in many processes, SPC provides a mechanism to generalize concerning process performance. SPC uses random samples generated consecutively over time. The random samples should be representative of the general process. SPC can be accomplished through the following steps:

- Control charts ($\bar{X}$-chart, R-chart)
- Process capability analysis (nested design, Cp, Cp_k);
- Process control (factorial design, response surface).

Control Charts

Two of the most commonly used control charts in industry are X-bar charts and range charts (R-charts). The type of chart to be used normally depends on the kind of data collected. Data collected can be of two types: variable data and attribute data. The success of quality improvement depends on two major factors:

1. The quality of data available

2. The effectiveness of the techniques used for analyzing the data

Types of Data for Control Charts

Variable data: The control charts for variable data are listed below:

- Control charts for individual data elements (X)
- Moving range chart (MR-chart)
- Average chart ($\bar{X}$-chart)
- Range chart (R-chart)
- Median chart
- Standard deviation chart (σ-chart)
- Cumulative sum chart (CUSUM)
- Exponentially weighted moving average (EWMA)

Attribute data: The control charts for attribute data are listed below:

- Proportion or fraction defective chart (p-chart) (subgroup sample size can vary)

- Percent defective chart (100p-chart) (subgroup sample size can vary)

- Number defective chart (np-chart) (subgroup sample size is constant)

- Number defective (c-chart) (subgroup sample size = 1)

- Defective per inspection unit (u-chart) (subgroup sample size can vary)

The statistical theory useful to generate control limits is the same for all the above charts with the exception of exponential weighted moving average (EWMA) and cumulative sum (CUSUM).

X-Bar and Range Charts

The R-chart is a time plot useful in monitoring short-term process variations, while the X-bar chart monitors the longer term variations where the likelihood of special causes is greater over time. Both charts have control lines called upper and lower control limits, as well as the central lines. The central line and control limits are calculated from the process measurements. They are not specification limits or a percentage of the specifications, or some other arbitrary lines based on experience. Therefore, they represent what the process is capable of doing when only common cause variation exists. If only common cause variation exists, then the data will continue to fall in a random fashion within the control limits. In this case, we say the process is in a state of statistical control. However, if a special cause acts on the process, one or more data points will be outside the control limits, so the process is not in a state of statistical control.

Data Collection Strategies

One strategy for data collection requires that about 20–25 subgroups be collected. 20 to 25 subgroups should adequately show the location and spread of a distribution in a state of statistical control. If it happens that due to sampling costs, or other sampling reasons associated with the process, we are unable to have 20–25 subgroups, we can still use the available samples that we have to generate the trial control limits and update these limits as more samples are made available, because these limits will normally be wider than normal control limits and will, therefore, be less sensitive to changes in the process. Another approach is to use run charts to monitor the process until such time as 20–25 subgroups are made available. Then, control charts can be applied with control limits included on the charts. Other data collection strategies should consider the subgroup sample size, as well as the sampling frequency.

Subgroup Sample Size

The subgroup samples of size n should be taken as n consecutive readings from the process and not random samples. This is necessary in order to have an accurate estimate of the process common cause variation. Each subgroup should be selected from some small period of time or small region of space or product in order to assure homogeneous conditions within the subgroup. This is necessary because the variation within the subgroup is used in generating the control limits. The subgroup sample size n can be between four or five samples. This is a good size that balances the pros and cons of using large or small sample size for a control chart as provided below.

Advantages of using small subgroup sample size

- Estimates of process standard deviation based on the range are as good and accurate as the estimates obtained from using the standard deviation equation which is a complex hand calculation method.

- The probability of introducing special cause variations within a subgroup is very small.

- Range chart calculation is simple and easier to compute by hand on the shop floor by operators.

Advantages of using large subgroup sample size

- The central limit theorem supports the fact that the process average will be more normally distributed with a larger sample size.

- If the process is stable, the larger the subgroup size the better the estimates of process variability.

- A control chart based on larger subgroup sample size will be more sensitive to process changes.

The choice of a proper subgroup is critical to the usefulness of any control chart. The following paragraphs explain the importance of subgroup characteristics:

- If we fail to incorporate all common cause variations within our subgroups, the process variation will be underestimated, leading to very tight control limits. Then the process will appear to go out of control too frequently even when there is no existence of a special cause.

- If we incorporate special causes within our subgroups, then we will fail to detect special causes as frequently as expected.

Frequency of Sampling

The problem of determining how frequently one should sample depends on several factors. These factors include but are not limited to the following:

- **Cost of collecting and testing samples**: The greater the cost of taking and testing samples, the less frequently we should sample.

- **Changes in process conditions**: The larger the frequency of changes to the process, the larger the sampling frequency. For example, if process conditions tend to change every 15 minutes, then sample every 15 minutes. If conditions change every two hours, then sample every two hours.

- **Importance of quality characteristics**: The more important the quality characteristic being charted is to the customer, the more frequently the characteristic will need to be sampled.

- **Process control and capability**: The more history of process control and capability, the less frequently the process needs to be sampled.

Stable Process

A process is said to be in a state of statistical control if the distribution of measurement data from the process has the same shape, location, and spread over time. In other words, a process is stable when the effects of all special causes have been removed from a process, so that the remaining variability is only due to common causes.

Out-of-Control Patterns

A process is said to be unstable (not in a state of statistical control) if it changes from time to time because of a shifting average, or shifting variability, or a combination of shifting averages and variation.

Calculation of Control Limits

- Range (R)

  This is the difference between the highest and lowest observations:

  $$R = X_{\text{highest}} - X_{\text{lowest}}$$

- Center lines

  Calculate $\bar{X}$ and $\bar{R}$

  $$\bar{X} = \frac{\sum X_i}{m}$$

  $$\bar{R} = \frac{\sum R_i}{m}$$

  where,

  $\bar{X}$ = overall process average

  $\bar{R}$ = average range

  m = total number of subgroups

  n = within subgroup sample size

- Control limits based on R-chart

  $UCL_R = D_4 \bar{R}$

  $LCL_R = D_3 \bar{R}$

- Estimate of process variation

  $$\hat{\sigma} = \frac{\bar{R}}{d_2}$$

- Control limits based on $\bar{X}$-chart

    Calculate the upper and lower control limits for the process average:

    $$UCL = \bar{X} + A_2\,\bar{R}$$
    $$LCL = \bar{X} - A_2\,\bar{R}$$

Plotting Control Charts for Range And Average Charts

- Plot the range chart (R-chart) first.

- If R-chart is in control, then plot X-bar chart.

- If R-chart is not in control, identify and eliminate special causes, then delete points that are due to special causes, and re-compute the control limits for the range chart. If the process is in control, then plot X-bar chart.

- Check to see if X-bar chart is in control, if not search for special causes and eliminate them permanently.

- Remember to perform the eight trend tests.

Plotting Control Charts for Moving Range And Individual Control Charts

- Plot the moving range chart (MR-chart) first.

- If MR-chart is in control, then plot the individual chart (X).

- If MR-chart is not in control, identify and eliminate special causes, then delete special causes points, and re-compute the control limits for the moving range chart. If MR-chart is in control, then plot the individual chart.

- Check to see if the individual chart is in control, if not search for special causes from out-of-control points.

- Perform the eight trend tests.

PROCESS CAPABILITY ANALYSIS FOR SIX SIGMA

Industrial process capability analysis is an important aspect of managing industrial projects. The capability of a process is the spread which contains almost all values of the process distribution. It is very important to note that capability is defined in terms of a distribution. Therefore, capability can only be defined for a process that is stable (has distribution) with common cause variation (inherent variability). It cannot be defined for an out-of-control process (which has no distribution) with variation special to specific causes (total variability).

Capable Process (C_p)

A process is capable ($C_p \geq 1$) if its natural tolerance lies within the engineering tolerance or specifications. The measure of process capability of a stable process is $6\hat{\sigma}$, where $\hat{\sigma}$ is the inherent process variability that is estimated from the process. A minimum value of $C_p = 1.33$ is generally used for an ongoing process. This ensures a very low rejection rate of 0.007% and therefore is an effective strategy for prevention of nonconforming items. C_p is defined mathematically as

$$C_p = \frac{USL - LSL}{6\hat{\sigma}}$$

$$= \frac{\text{allowable process spread}}{\text{actual process spread}}$$

where:
    USL = upper specification limit
    LSL = lower specification limit

C_p measures the effect of the inherent variability only. The analyst should use R-bar/d_2 to estimate $\hat{\sigma}$ from an R-chart that is in a state of statistical control, where R-bar is the average of the subgroup ranges and d_2 is a normalizing factor that is tabulated for different subgroup sizes (n). We don't have to verify control before performing a capability study. We can perform the study, then verify control after the study with the use of control charts. If the process is in control during the study, then our estimates of capabilities are correct and valid. However, if the process was not in control, we would have gained useful information, as well as proper insights as to the corrective actions to pursue.

Capability Index (C_{pk})

Process centering can be assessed when a two-sided specification is available. If the capability index (C_{pk}) is equal to or greater than 1.33, then the process may be adequately centered. C_{pk} can also be employed when there is only one-sided specification. For a two-sided specification, it can be mathematically defined as:

$$C_{pk} = \text{Minimum} \left\{ \frac{USL - \bar{X}}{3\hat{\sigma}}, \frac{\bar{X} - LSL}{3\hat{\sigma}} \right\}$$

where:

$\bar{X}$ = Overall process average

However, for a one-sided specification, the actual C_{pk} obtained is reported. This can be used to determine the percentage of observations out of specification. The overall long-term objective is to make C_p and C_{pk} as large as possible by continuously improving or reducing process variability, $\hat{\sigma}$, for every iteration so that a greater percentage of the product is near the key quality characteristics target value. The ideal is to center the process with zero variability.

If a process is centered but not capable, one or several courses of action may be necessary. One of the actions may be that of integrating designed experiment to gain additional knowledge on the process and in designing control strategies. If excessive variability is demonstrated, one may conduct a nested design with the objective of estimating the various sources of variability. These sources of variability can then be evaluated to determine what strategies to use in order to reduce or permanently eliminate them. Another action may be that of changing the specifications or continuing production and then sorting the items. Three characteristics of a process can be observed with respect to capability, as summarized in the numbered list.

1. The process may be centered and capable.

2. The process may be capable but not centered.

3. The process may be centered but not capable.

Process capability example

Step 1: Using data for the specific process, determine if the process is capable. Let us assume that the analyst has determined that the process is in a state of statistical control. For this example, the specification limits are set at 0 (lower limit) and 45 (upper limit). The inherent process variability as determined from the control chart is:

$$\hat{\sigma} = \bar{R} / d_2 = 5.83 / 2.059 = 2.83$$

The capability of this process to produce within the specifications can be determined as:

$$C_p = \frac{USL - LSL}{6\hat{\sigma}} = \frac{45 - 0}{6(2.83)} = 2.650.$$

The capability of the process, $C_p = 2.65 > 1.0$ indicating that the process is capable of producing clutches that will meet the specifications of between 0 and 45. The process average is 29.367.

Step 2: Determine if the process can be adequately centered. C_{pk} = minimum [C_l and C_u] can be used to determine if a process can be centered.

$$C_u = \frac{USL - \bar{X}}{3\hat{\sigma}} = \frac{45 - 29.367}{3(2.83)} = 1.84$$

$$C_l = \frac{\bar{X} - LSL}{3\hat{\sigma}} = \frac{29.367 - 0}{3(2.83)} = 3.46$$

Therefore, the capability index, C_{pk}, for this process is 1.84. Since C_{pk} = 1.84 is greater than 1.33, then the process can be adequately centered.

LEAN PRINCIPLES AND APPLICATIONS

What is "Lean"? Lean means the identification and elimination of sources of *waste* in operations. Recall that Six Sigma involves the identification and elimination of the source of *defects*. When Lean and Six Sigma are coupled, an organization can derive the double benefit of reducing waste and defects in operations; which leads to what is known as Lean Six Sigma. Consequently, the organization can achieve higher product quality, better employee morale, better satisfaction of customer requirements, and more effective utilization of limited resources. The basic principle of "lean" is to take a close look at the elemental compositions of a process so that non-value-adding elements can be located and eliminated.

APPLYING KAIZEN TO A PROCESS

By applying the Japanese concept of "*Kaizen*," which means "take apart and make better," an organization can redesign its processes to be lean and devoid of excesses. In a mechanical design sense, this can be likened to finite element analysis, which identifies how the component parts of a mechanical system fit together. It is by identifying these basic elements that improvement opportunities can be easily and quickly recognized. It should be recalled that the process of work breakdown structure in project management facilitates the identification of task-level components of an endeavor. Consequently, using a project management approach facilitates the achievement of the objectives of "lean." The functional relationships (f) are summarized as shown below:

Task = f(activity)

Sub-process = f(task)

Process = f(sub-process)

Quality system = f(process)

Thus, quality improvement can be achieved by hierarchically improving a process and all the elements contained therein. Fads come and go in industry. Over the years, we have witnessed the introduction and demise of many techniques that were hailed as the panacea of industry's ailments. Some of the techniques have survived the test of time because they do, indeed, hold some promise. Lean techniques appear to hold such promise, if it is viewed as an open-ended but focused application of the many improvement tools that have emerged over the years. The adoption of lean principles by the US Air Force has given more credence to its application. The US Air Force embarked on a massive endeavor to achieve widespread improvement in operational processes throughout the Air Force. The endeavor is called AFSO21 (Air Force Smart Operations for the 21st Century or simply Air Force Smart Ops 21). This endeavor requires the implementation of appropriate project management practices at all levels. AFSO21 is a coordinated effort at achieving operational improvement in US Air Force operations; throughout the rank and file of the large organization. It is an integrative process of using Lean Principles, Theory of Constraints, Six Sigma, BPI, MBO, TQM, 6s, Project Management, and other classical management tools. However, the implementation of Lean principles constitutes about 80 percent of AFSO21 efforts. As a part of tools for lean practices and procedures, the following section presents lean task value rating system, which helps to compare and rank elements of a process for retention, re-scoping, scaling, or elimination purposes.

Lean Six-Sigma Within Project Management

Lean and Six Sigma use analytical tools as the basis for pursuing their goals. But the achievement of those goals is predicated on having a structured approach to the activities of production. If proper project management is practiced at the outset in an industrial endeavor, it will pave the way for achieving Six Sigma results and realizing lean outcomes. The key in any project endeavor is to have a structured design of the project so that diagnostic and corrective steps can easily be pursued. If the proverbial "garbage" is allowed to creep into a project, it would take much more time, effort, and cost to achieve a Lean Six Sigma cleanup.

REFERENCES

Badiru, A. B. and B. J. Ayeni, *Practitioner's Guide to Quality and Process Improvement*, Chapman & Hall, London, 1993.

PMI, *A Guide to the Project Management Body of Knowledge (PMBOK Guide)*, 6th Edition, Project Management Institute, Philadelphia, PA, 2017.

CHAPTER 6

Human Resource Management

P ROJECT HUMAN RESOURCE MANAGEMENT is the next stage of the structural approach to project management in PMBOK guidelines (PMI, 2017; Sawhney, Badiru and Niranjan, 2004). It is a common practice to ask for additional human resources. But care must be exercised to determine where and when additional human resources are really needed and how they are deployed. A pertinent question to ask is: What incremental value or benefit is provided by the addition of one more unit of resource? This is of particular interest when allocating technical human resources in all projects.

Human resource management provides the foundation for accomplishing project goals. Even in highly automated environments, human resources are still a key element in accomplishing goals and objectives. Human resource management involves the function of directing human resources throughout a project's life cycle. This requires the art and science of behavioral knowledge to achieve project goals. Employee involvement and empowerment are crucial elements of achieving the quality expectations of a project. The project manager is the key player in human resource management. Good leadership qualities and inter-personal skills are essential for dealing with both internal and external human resources associated with a project. The legal and safety aspects of employee welfare are important factors in human resource management. Human resource management is carried out to express and uphold an organization's standards and expectations in terms of the attributes of employees including the following:

- Desired competencies
- Current assignments
- Performance metrics
- Development plans
- Outline of future goals

Standard competencies of employees must be factored into in-house and customized reviews for project teams of technical professionals. Employee performance reviews should be tailored to employees' functional positions and accountability. Some standard competencies include the following:

- Achievement of focus
- Business and operational acumen
- Consultative personality
- Work design ability
- Continuous learning
- Being customer focus
- Quality management
- Strategic thinking

- Team leadership
- Technology awareness
- Vision

Designing human systems interaction into a STEP project environment ensures that human resources can integrate well with sophisticated science, technology, and engineering tools. Issues such as human effectiveness, performance excellence, occupational health, readiness, and safety are essential components of planning STEP projects from a human resource management perspective.

AGING WORKFORCE IN SCIENCE, TECHNOLOGY, AND ENGINEERING

The issue of the aging workforce in science, technology, and engineering is very critical. Scientists and engineers that form the backbone of the advancements during and post–World War II are gradually phasing out of the STEP landscape. Managing and advancing contemporary science, technology, and engineering projects requires an infusion of a new crop of researchers, educators, and practitioners in those fields. Many nations are scrambling to find ways to train, acquire, or retain qualified STE professionals. Not only must the desired replenishment workforce be developed, it must also be preserved through the following strategies:

- Challenge the workforce to engage in higher-level educational opportunities.

- Provide smooth avenues for the younger generation to follow the path of science, technology, and engineering.

- Create cohesive and progressive linkage between elementary education and higher education (e.g., K-12 to graduate studies education initiatives).

- Recognizing that science and technology are subject to dynamism and job uncertainties, create a guarantee of job security.

- Recognizing the high value of science, technology, and engineering professionals, institute programs that assure personal safety and security of the workforce.

- Recognizing that longevity is essential for getting the most out of the STE workforce, provide healthcare programs that extend and preserve the services of the workforce for as long as practicable.

- Diversify local availability of STE jobs to encourage recruitment, relocation, and local retention of the workforce.

- Encourage localities for provide physical infrastructure and favorable operating conditions to attract and retain science and technology industry.

- Create avenues for expanding job opportunities for the new and younger workforce so that "new" and "old" can coexist productively.

- Create tiered and mentoring workforce relationship such that age discrimination does not creep into workforce relationships.

- Take advantage of the wisdom and experience of the outgoing (retiring and departing) workforce in preparing the incoming (next generation) workforce.

KNOWLEDGE WORKERS IN TECHNICAL WORK ENVIRONMENT

Knowledge workers will be the boon of the technical work environment of the future. While the brawn of yesteryears will still be needed in some quarters of the economy, the brain of the future is what will be needed to achieve and sustain technological advantage. A major shift will have to be instituted to accommodate the needs of knowledge workers. While work–time accountability will still be needed for compensation purposes, the major shift will be to judge knowledge workers on the basis of their accomplishments rather than the number of hours spent directly on the job. Knowledge workers often work around the clock, even when they are not at work. Higher-level strategy formulation, rationalization, and pensive reflection on work actions most often occur away from the hustle and bustle of the actual work environment. This off-site brain-intensive "work" translates directly or indirectly into work-site accomplishments. Many organizations already recognize this benefit and are already accounting for overall accomplishments in their compensation packages for employees. But this practice needs to spread to more organizations throughout the rank and file of the economy; even in industries that are not traditionally seen as being of high-tech caliber. The following phrase re-emphasizes the point of the unique asset that knowledge workers bring to the work environment:

Evaluate knowledge workers on the basis of their overall accomplishments rather than how many office-hours it takes to achieve the accomplishments.

ELEMENTS OF HUMAN RESOURCE MANAGEMENT

Human resources are the basis of managing projects. Even highly automated systems must have human intervention at specific points to ensure overall process efficiency. Inter-human relationships must, thus, play a major role in an organization's strategy for managing human resources. Human divides don't mend easily. An organization must work hard to prevent a divide in the first place. This requires every organization to recognize and interface the following three major components of a project:

1. People (Managers, Team members, Stockholders, Vendors, Suppliers, etc.)

2. Processes (Work design, Lean initiative, Six Sigma, Business process re-engineering, etc.)

3. Tools (Technology, Widgets, Facilities, Information, etc.)

The success of any project is dependent on the human resources associated with linking its components. Human resources are distinguished from other resources because of the ability to learn, adapt to new project situations, and set goals. Human resources, technology resources, and management resources must coexist to pursue project goals. Managing human resources involves placing the right people with the right skills in the right jobs in the right environment. Good human resource management motivates workers to perform better. Both individual and organizational improvements are needed to improve overall quality by enriching jobs with the following strategies:

- Specify project goals in unambiguous terms

- Encourage and reward creativity on the job

- Eliminate mundane job control processes

- Increase accountability and responsibility for project results

- Define jobs in terms of manageable work packages that help identify line of responsibility

- Grant formal authority to make decisions at the task level

- Create advancement opportunities in each job

- Give challenging assignments that enable a worker to demonstrate his/her skill

- Encourage upward (vertical) communication of ideas

- Provide training and tools needed to get job done

- Maintain a stable management team

Several management approaches are used to manage human resources. Some of these approaches are formulated as direct responses to the cultural, social, family, or religious needs of workers. Examples of these approaches are:

- Flextime
- Religious holidays
- Half-time employment

These approaches can have a combination of several advantages. Some of the advantages are for the employer, while some are for the workers. The advantages are presented below:

- Low cost
- Cost savings on personnel benefits
- Higher employee productivity
- Less absenteeism
- Less work stress
- Better family/domestic situation, which may have positive effects on productivity

Work force retraining is important for automation projects. Continuing education programs should be developed to retrain people who are only qualified to do jobs that do not require skilled manpower. The retraining will create a ready pool of human resource that can help boost manufacturing output and competitiveness. Management stability is needed to encourage workers to adapt to the changes in industry. If management changes too often, workers may not develop a sense of commitment to the policies of management.

The major resource in any organization is manpower both technical and non-technical. People are the overriding factor in any project life cycle. Even in automated operations, the role played by whatever few people are involved can be very significant. Such operations invariably require the services of technical people with special managerial and professional needs. The high-tech manager in such situations would need special skills in order to discharge the managerial duties effectively. The manager must have auto-management skills that relate to the following:

- Managing self

- Being managed

- Managing others

Many of the managers who supervise technical people rise to the managerial posts from technical positions. Consequently, they often lack the managerial competence needed for the higher offices. In some cases, technical professionals are promoted to managerial levels and then transferred to administrative posts in functional areas different from their areas of technical competence. The poor managerial performance of these technical managers is not necessarily a reflection of poor managerial competence, but rather an indication of the lack of knowledge of the work elements in their surrogate function. Any technical training without some management exposure is, in effect, an incomplete education. Technical professionals should be trained for the eventualities of their professions.

In the transition from the technical to the management level, an individual's attention would shift from detail to overview, specific to general, and technical to administrative. Since most managerial positions are earned based on qualifications (except in aristocratic and autocratic systems), it is important to train technical professionals for possible administrative jobs. It is the responsibilities of the individual and the training institution to map out career goals and paths and institute specific education aimed at the realization of those goals. One such path is outlined below.

1. **Technical professional**: This is an individual with practical and technical training and/or experience in a given field, such as industrial engineering. The individual must keep current in his/her area of specialization through continuing education courses, seminars, conferences, and so on. The mentor program, which is now used in many large organizations, can be effectively utilized at this stage of the career ladder.

2. **Project manager**: This is an individual assigned the direct responsibility of supervising a given project through the phases of planning, organizing, scheduling, monitoring, and control. The managerial assignment may be limited to just a specific project. At the conclusion of the project, the individual returns to his/her regular technical duties. However, his/her performance on the project may help identify him/her as a suitable candidate for permanent managerial assignment later on.

3. **Group manager**: This is an individual who is assigned direct responsibility to plan, organize, and direct the activities of a group of people with a specific responsibility—for example, a computer data security advisory committee. This is an ongoing responsibility that may repeatedly require the managerial skills of the individual.

4. **Director**: An individual who oversees a particular function of the organization. For example, a marketing director has the responsibility of developing and implementing the strategy for getting the organization's products to the right market, at the right time, at the appropriate price, and in

the proper quantity. This is a critical responsibility that may directly affect the survival of the organization. Only the individuals who have successfully proven themselves at the earlier career stages get the opportunity to advance to the director's level.

5. **Administrative manager**: This is an individual who oversees the administrative functions and staff of the organization. His/her responsibilities cut across several functional areas. He/she must have proven his/her managerial skills and diversity in previous assignment.

The previous is just one of the several possible paths that can be charted for a technical professional as he/she gradually makes the transition from the technical ranks to the management level. To function effectively, a manager must acquire a non-technical background in various subjects. His/her experience, attitude, personality, and training will determine his/her managerial style. His/her appreciation of the human and professional needs of his subordinates will substantially enhance his/her managerial performance. Examples of subject areas in which a manager or an aspiring manager should get training include the ones outlined below.

1. Project management

   a. *Scheduling and budgeting*: Knowledge of project planning, organizing, scheduling, monitoring, and controlling under resource and budget restrictions.

   b. *Supervision*: Skill in planning, directing, and controlling the activities of subordinates.

   c. *Communication*: Skill of relating to others both within and outside the organization. This includes written and oral communication skills.

2. Personal and Personnel Management

a. *Professional development*: Leadership roles played by participating in professional societies and peer recognition acquired through professional services.

b. *Personnel development*: Skills needed to foster cooperation and encouragement of staff with respect to success, growth, and career advancement.

c. *Performance evaluation*: Development of techniques for measuring, evaluating, and improving employee performance.

d. *Time management*: Ability to prioritize and delegate activities as appropriate to maximize accomplishments within given time periods.

3. Operations management

a. *Marketing*: Skills useful for winning new business for the organization or preserving existing market shares.

b. *Negotiating*: Skills for moderating personnel issues, representing the organization in external negotiations, or administering company policies.

c. *Estimating and budgeting*: Skills needed to develop reasonable cost estimates for company activities and the assignment of adequate resources to operations.

d. *Cash Flow analysis*: An appreciation for the time value of money, manipulations of equity and borrowed capitals, stable balance between revenues and expenditures, and maximization of returns on investments.

e. *Decision analysis*: Ability to choose the direction of work by analyzing feasible alternatives.

A technical manager can develop the above skills through formal college courses, seminars, workshops, short courses, professional conferences, or in-plant company training. Several companies appreciate the need for these skills and are willing to bear the cost of furnishing their employees with the means of acquiring the skills. Many of the companies have custom formal courses which they contract out to colleges to teach for their employees. This is a unique opportunity for technical professionals to acquire managerial skills needed to move up the company ladder.

Technical people have special needs. Unfortunately, some of these needs are often not recognized by peers, superiors, or subordinates. Inexperienced managers are particularly prone to the mistake of not distinguishing between technical and nontechnical professional needs. In order to perform more effectively, a manager must be administratively adaptive. He/she must understand the unique expectations of technical professionals in terms of professional preservation, professional peers, work content, hierarchy of needs, and the technical competence or background of their managers. Maslow's hierarchy of needs presents the organizational theory of how individual workers behave and respond to stimuli in the project environment.

Maslow's Hierarchy of Needs covers the basic elements summarized below:

1. **Physiological Needs**: The needs for the basic necessities of life, such as food, water, housing, and clothing (Survival Needs). This is the level where access to wages is most critical. Biological needs fall into this category also.

2. **Safety Needs**: The needs for security, stability, and freedom from threat of physical harm. Desire for safe working environment.

3. **Social Needs**: The needs for social acceptance (sense of belonging), friends, love, affection, and association. Industrial outsourcing may bring about better economic

outlook that may enable each individual to be in a better position to meet his or her social needs.

4. **Esteem Needs**: The needs for accomplishment, respect, recognition, attention, self-respect, autonomy, and appreciation. These needs are important not only at the individual level, but also at the organizational level.

5. **Self-Actualization Needs**: These are the needs for self-fulfillment and self-improvement. They also involve the stage of opportunity to grow professionally. Industrial outsourcing may create opportunities for individuals to assert themselves socially and economically.

In addition to Abraham Maslow's hierarchy of needs, the following motivation theories are also essential:

- Theory X and Theory Y (presented by Douglas McGregor)
- Motivation-Hygiene Factors (presented by Frederick Herzberg)

Theory X can be effectively utilized in developing project teams. It has the following doctrines:

- Workers inherently dislike work, and whenever possible, will attempt to avoid it.
- Since workers dislike work, they must be coerced, controlled, cajoled, coaxed, enticed, persuaded, or threatened with punishment to achieve desired goals.
- Workers will evade and shift responsibilities and seek formal direction whenever possible.
- Most workers place security above all other factors associated with work and will display little ambition or self-motivation.

Theory Y can be embraced to take advantage of positive self-direction and self-actuating nature of workers to achieve project goals with little or no external prompting. The basic doctrines of Theory Y are:

- Workers can view work as being as natural as personal normal pursuits such as recreation.

- A worker who is committed to the objectives of a project will exhibit self-direction, self-actuation, and self-control to get the job done.

- The average worker readily accepts and even seeks responsibilities to get the job done.

- Creativity, the ability to make good decisions, permeates the organization; and is not necessarily limited to the select few in management.

Motivation-Hygiene factors also present constructive attributes of work that facilitate the achievement of project objectives.

- Motivation factors include achievement, recognition, job growth, work design itself, increased responsibility, career advancement.

- Hygiene factors include company policy, supervision practices, interpersonal relationships, work conditions, salary compensation, bonus program, personal life, professional status, and job security.

Professional Preservation. Professional preservation refers to the desire of a technical professional to preserve his/her identification with a particular job function. In many situations, the preservation is not possible due to a lack of manpower to fill specific job slots. It is common to find people trained in one technical field holding assignments in other fields. An incompatible job function can easily become the basis for insubordination, egotism, and rebellious attitudes. While it is realized that in any job environment there will sometimes be the need to work outside one's profession, every effort should be made to match the surrogate profession as close as possible. This is primarily the responsibility of the human resources manager.

After a personnel team has been selected in the best possible manner, a critical study of the job assignments should be made. Even between two dissimilar professions, there may be specific job functions that are compatible. These should be identified and used in the process of personnel assignment. In fact, the mapping of job functions needed for an operation can serve as the basis for selecting a project team. In order to preserve the professional background of technical workers, their individualism must be understood. In most technical training programs, the professional is taught how to operate in the following ways:

1. Make decisions based on the assumption of certainty of information

2. Develop abstract models to study the problem being addressed

3. Work on tasks or assignments individually.

4. Quantify outcomes

5. Pay attention to exacting details

6. Think autonomously

7. Generate creative insights to problems

8. Analyze systems operability rather than profitability

However, in the business environment, not all of the above characteristics are desirable or even possible. For example, many business decisions are made with incomplete data. In many situations, it is unprofitable to expend the time and efforts to seek perfect data. As another example, many operating procedures are guided by company policies rather than the creative choices of employees. An effective manager should be able to spot cases where a technical employee may be given room to practice his professional training. The job design should be such that the employee can address problems in a manner compatible with his professional training.

Professional Peers. In addition to having professionally compatible job functions, technical people like to have other project team members to whom they can relate technically. A project team consisting of members from diversely unrelated technical fields can be a source of miscommunication, imposition, or introversion. The lack of a professional associate on the same project can cause a technical person to exhibit one or more of the following attitudes:

1. Withdraw into a shell; and contribute very little to the project by holding back ideas that he/she feels the other project members cannot appreciate

2. Exhibit technical snobbery; and hold the impression that only he/she has the know-how for certain problems

3. Straddle the fence on critical issues; and develop no strong conviction for project decisions

Providing an avenue for a technical "buddy system" to operate in an organization can be very instrumental in ensuring congeniality of personnel teams and in facilitating the eventual success of project endeavors. The manager in conjunction with the selection committee (if one is used) must carefully consider the mix of the personnel team on a given project. If it is not possible or desirable to have more than one person from the same technical area on the project, an effort should be made to provide as good a mix as possible. It is undesirable to have several people from the same department taking issues against the views of a lone project member from a rival department. Whether it is realized or not, whether it is admitted or not, there is a keen sense of rivalry among technical fields. Even within the same field, there are subtle rivalries between specific functions. It is important not to let these differences carry over to a project environment.

Work Content. With the advent of new technology, the elements of a project task will need to be designed to take advantage of new developments. Technical professionals have a sense of achievement relative to their expected job functions. They will not be satisfied with mundane project assignments that will bring forth their technical competence. They prefer to claim contribution mostly where technical contribution can be identified. The project manager will need to ensure that the technical people of a project have assignments for which their background is really needed, it will be counterproductive to select a technical professional for a project mainly on the basis of personality. An objective selection and appropriate assignment of tasks will alleviate potential motivational problems that could develop later in the project.

Hierarchy of Needs. Recalling Maslow's hierarchy of needs, the needs of a technical professional should be more critically analyzed. Being professionals, technical people are more likely to be higher up in the needs hierarchy. Most of their basic necessities for a good life would already have been met. Their prevailing needs will tend to involve esteem and self-actualization. As a result, by serving on a project team, a technical professional may have expectations that cannot usually be quantified in monetary terms. This is in contrast to non-technical people who may look forward to overtime pay or other monetary gains that may result from being on the project. Technical professionals will generally look forward to one or several of the following opportunities.

1. *Professional growth and advancement*: Professional growth is a primary pursuit of most technical people. For example, a computer professional has to be frequently exposed to

challenging situations that introduce new technology developments and enable him to keep abreast of his field. Even occasional drifts from the field may lead to the fear of not keeping up and being left behind. The project environment must be reassuring to the technical people with regard to the opportunities for professional growth in terms of developing new skills and abilities.

2. *Technical freedom*: Technical freedom, to the extent permissible within the organization, is essential for the full utilization of a technical background. A technical professional will expect to have the liberty of determining how best the objective of his assignment can be accomplished. One should never impose a work method on a technical professional with the assurance that "this is the way it has always been done and will continue to be done!" If the worker's creative input to the project effort is not needed, then there is no need to have him or her on the team in the first place.

3. *Respect for personal qualities*: Technical people have profound personal feelings despite the mechanical or abstract nature of their job functions. They will expect to be respected for their personal qualities. In spite of frequently operating in professional isolation, they do engage in interpersonal activities. They want their non-technical views and ideas to be recognized and evaluated based on merit. They don't want to be viewed as "all technical." An appreciation for their personal qualities gives them the sense of belonging and helps them to become productive members of a project team.

4. *Respect for professional qualification*: A professional qualification usually takes several years to achieve and is not likely to be compromised by any technical professional. Technical professionals cherish the attention they receive due to their technical background. They expect certain preferential

treatments. They like to make meaningful contributions to the decision process. They take approval of their technical approaches for granted. They believe they are on a project because they are qualified to be there. The project manager should recognize these situations and avoid the bias of viewing the technical person as being conceited.

5. *Increased recognition*: Increased recognition is expected as a by-product of a project effort. The technical professional, consciously or subconsciously, views his participation in a project as a means of satisfying one of his higher-level needs. He/she expects to be praised for the success of his/her efforts. He/she looks forward to being invited for subsequent technical endeavors. He/she savors hearing the importance of his contribution being related to his/her peers. Without going to the extreme, the project manager can ensure the realization of the above needs through careful comments.

6. *New and rewarding professional relationships*: New and rewarding professional relationships can serve as a bonus for a project effort. Most technical developments result from the joint efforts of people that share closely allied interests. Professional allies are most easily found through project groups. A true technical professional will expect to meet new people with whom he/she can exchange views, ideas, and information later on. The project atmosphere should, as a result, be designed to be conducive to professional interactions.

Quality of Leadership. The professional background of the project leader should be such that he/she commands the respect of technical subordinates. The leader must be reasonably conversant with the base technologies involved in the project. He/she must be able to converse intelligently on the terminologies of the project topic and be able to convey the project ideas to upper management. This serves to give him/her technical credibility. If technical credibility is lacking, the technical professionals on the project might view him/her as an ineffective leader. They will consider it impossible to serve under a manager to whom they cannot relate technically.

In addition to technical credibility, the manager must also possess administrative credibility. There are routine administrative matters that are needed to ensure a smooth progress for the project. Technical professionals will prefer to have those administrative issues successfully resolved by the project leader so that they can concentrate their efforts on the technical aspects. The essential elements of managing a group of technical professionals involve identifying the unique characteristics and needs of the group and then developing the means of satisfying those unique needs.

Recognizing the peculiar characteristics of technical professionals is one of the first steps in simplifying project management functions. The nature of manufacturing and automation projects calls for the involvement of technical human resources. Every manager must appreciate the fact that the cooperation or the lack of cooperation from technical professionals can have a significant effect on the overall management process. The success of a project can be enhanced or impeded by the management style utilized.

Work Simplification. Work simplification is the systematic investigation and analysis of planned and existing work systems and methods for the purpose of developing easier, quicker, less fatiguing, and more economic ways of generating high-quality goods and services. Work simplification facilitates the content of workers, which invariably leads to better performance. Consideration must be given to improving the product or service, raw materials and supplies, the sequence of operations, tools, work place, equipment, and hand and body motions. Work simplification analysis helps in defining, analyzing, and documenting work methods.

Human Resource Management: Step-by-Step Implementation

The human resource management component of the project management body of knowledge consists of the elements shown below. The four elements are carried out across the process groups. The overlay of the elements and the process groups are shown in Table 6.1. Thus, under the knowledge area of human resource management, the required steps are:

Step 1: Human Resource Planning

Step 2: Acquire Project Team

Step 3: Develop Project Team

Step 4: Manage Project Team

Human resource planning involves identifying and documenting project roles, responsibilities, and creating staffing management plan. Availability, cost, and competence are essential attributes to be evaluated at this stage. Acquire project team involves obtaining the human resources needed to complete the project. Acquiring human resources involves negotiation as well as the possibility of using virtual teams. Developing the project team involves improving the

TABLE 6.1 Implementation of Human Resource Management Across Process Groups

Initiating
Planning
   1. Human Resource Planning
Executing
   1. Acquire Project Team
   2. Develop Project Team
Monitoring and Controlling
   1. Manage Project Team
Closing

competencies and interactions of team members to enhance project performance. Developing human resources requires effective communication, motivation, problem solving, work facilitation, and influencing. Managing the project team involves tracking team member performance, providing feedback, resolving conflicts, and coordinating changes to enhance project performance. Managing human resources implies team building and implementing project team ground rules. Tables 6.2 through 6.5 present the inputs, tools, techniques, and outputs of each step.

TABLE 6.2   Tools and Techniques for Human Resource Planning within Project Human Resource Management

STEP 1: Human Resource Planning

Inputs

Enterprise environmental factors
Organizational process assets
Project management plan
Activity resource requirements
Other in-house (custom) factors of relevance and interest

Tools and Techniques

Organizational charts
Team networking
Group dynamics
Organizational theory
Other in-house (custom) tools and techniques

Output(s)

Roles and responsibilities
Project organization charts
Staffing outline
Management plan
Other in-house outputs, reports, and data inferences of interest to the
  organization

TABLE 6.3   Tools and Techniques for Acquiring Project Team within Project Human Resource Management

STEP 2: Acquire Project Team

Inputs

Enterprise environmental factors
Organizational process assets
Roles and responsibilities
Project organization charts
Staffing management plan
Other in-house (custom) factors of relevance and interest

Tools and Techniques

Pre-assignment
Negotiation
Acquisition
Virtual teaming
Staff exchange programs
Co-location programs
Other in-house (custom) tools and techniques

Output(s)

Project staff assignments
Resource availability database
Staffing management plan (updates)
Other in-house outputs, reports, and data inferences of interest to the organization

MANAGING HUMAN RESOURCE PERFORMANCE

Connecting with the Employee is a basic requirement of managing human resource performance. Unbiased leadership means not judging others based on one's own values. As a leader, a project manager cannot be a good communicator if he/she is not a good listener. The LINK (Look, Inquire, Note, Know) concept for connecting with employees requires that the leader exhibit empathy for the employee's specific needs. Managing human resource performance requires the following:

TABLE 6.4 Tools and Techniques for Developing Project Team within Project Human Resource Management

STEP 3: Develop Human Resource

Inputs
Project staff assignments
Staffing management plan
Resource availability
Other in-house (custom) factors of relevance and interest

Tools and Techniques
General management skills
Training
Team-building exercises
Ground rules formulation
Co-location strategies
Recognition and awards
Other in-house (custom) tools and techniques

Output(s)
Team performance assessment
Other in-house outputs, reports, and data inferences of interest to the organization

1. Managing employee information

2. Setting and managing goals

3. Documenting ongoing performance events

4. Developing employee performance improvement and advancement and strategies

Negligence in managing employee information will most certainly lead to inaccuracies that will adversely affect how effectively performance can be managed or rectified.

TABLE 6.5   Tools and Techniques for Managing Project Team within Project
Human Resource Management

STEP 4: Manage Human Resource

Inputs

Organizational process assets

Project staff assignments

Roles and responsibilities

Project organization chart

Staffing management plan

Team performance assessment

Performance reports

Other in-house (custom) factors of relevance and interest

Tools and Techniques

Triple C Model (communication, cooperation, coordination)

Hierarchy of Needs

Theory X and Theory Y

Motivation-hygiene factors

Management by objective

Management by exception

Observational programs

Staff conversation and dialogue techniques

Conflict management

Issue log

Other in-house (custom) tools and techniques

Output(s)

Requested changes

Record of corrective actions

Record of preventive actions

Organizational process assets

Project management plan (updates)

Other in-house outputs, reports, and data inferences of interest to the
organization

REFERENCES

PMI, *A Guide to the Project Management Body of Knowledge (PMBOK Guide)*, 6th Edition, Project Management Institute, Philadelphia, PA, 2017.

Sawhney, R.; A. B. Badiru; and A. Niranjan, "A Model for Integrating and Managing Resources for Technical Training Programs," in *Internet Economy: Opportunities and Challenges for Developed and Developing Regions of the World*, Y. A. Hosni and T. M. Khalil, editors, Elsevier, Boston, MA, 2004, pp. 337–351.

CHAPTER 7

Communications Management

"Communication is the root of everything else"

ADEDEJI BADIRU

As the original quote in the chapter title suggests, communication is vital to everything else in a project. Any successful project manager should spend 90% of his or her time on communications activities. This is a vital function that is even more crucial in technology-based projects. Communications management refers to the functional interface between individuals and groups within the project environment (PMI, 2017; Mooz, Forsberg and Cotterman, 1993). This involves proper organization, routing, and control of information needed to facilitate work. Good communication is in effect when there is a common understanding of information between the communicator and the target. Communications management facilitates unity of purpose in the project environment. The success of a project is directly related to the effectiveness of project communication. From the author's experience, most project problems can be traced to a lack of proper communication. Communication is achieved through a variety of means beyond verbal exchanges. Telling, showing, and direct involvement are all effective modes of communication. A Chinese proverb says,

> Tell me, and I forget;
>
> Show me, and I remember;
>
> Involve me, and I understand.

The project team should employ all possible avenues to get project information across to everyone.

COMMUNICATIONS MANAGEMENT: STEP-BY-STEP IMPLEMENTATION

The four elements in the communication block are carried out across the process groups. The overlay of the elements and the process groups are shown in Table 7.1. Thus, under the knowledge area of communications management, the required steps are:

Step 1: Communications Planning

Step 2: Information Distribution

Step 3: Performance Reporting

Step 4: Manage Stakeholders

TABLE 7.1   Implementation of Communications Management Across Process Groups

Initiating
Planning
   1. Communications Planning
Executing
   1. Information Distribution
Monitoring and Controlling
   1. Performance Reporting
   2. Manage stakeholders
Closing

Tables 7.2 through 7.5 present the inputs, tools, techniques, and outputs of each step of communications management. Communications planning involves determining the information and communication needs of the stakeholders regarding who needs what information, when, where, and how. Information distribution involves making the needed information available to project stakeholders in a timely manner and in an appropriate dosage. Performance reporting involves collecting and disseminating performance information, which includes status reporting, progress measurement, and forecasting. Managing stakeholders involves managing communications to satisfy the requirements of the stakeholders so as to resolve issues that develop.

COMPLEXITY OF MULTI-PERSON COMMUNICATION

Communication complexity increases with an increase in the number of communication channels. It is one thing to wish to

TABLE 7.2   Tools and Techniques for Communications Planning within Project Communications Management

STEP 1: Communications Planning

Inputs

Enterprise environmental factors

Organizational process assets

Project scope statement

Project constraints and assumptions

Other in-house (custom) factors of relevance and interest

Tools and Techniques

Communications requirement analysis

Communications technology

Communications responsibility matrix

Collaborative alliance

Other in-house (custom) tools and techniques

Output(s)

Communications

Management plan

Other in-house outputs, reports, and data inferences of interest to the
  organization

TABLE 7.3   Tools and Techniques for Information Distribution within Project Communications Management

STEP 2: Information Distribution

Inputs

Communication management plan

Personnel distribution list

Other in-house (custom) factors of relevance and interest

Tools and Techniques

Communication modes and skills

Social networking

Influence networking

Meetings and dialogues

Communication relationships

Information gathering and retrieval systems

Information distribution methods

Lessons learned

Best practices

Information exchange

Other in-house (custom) tools and techniques

Output(s)

Organizational process assets (updates)

Other in-house outputs, reports, and data inferences of interest to the
  organization

communicate freely, but it is another thing to contend with the increased complexity when more people are involved. The statistical formula of combination can be used to estimate the complexity of communication as a function of the number of communication channels or number of participants. The combination formula is used to calculate the number of possible combinations of r objects from a set of n objects. This is written as:

$$_nC_r = \frac{n!}{r!\left[n-r\right]!}$$

In the case of communication, for illustration purposes, we assume communication is between two members of a team at a

TABLE 7.4   Tools and Techniques for Performance Reporting within Project
Communications Management

STEP 3: Performance Reporting

Inputs
Work performance information
Performance measurements
Forecasted completion
Quality control measurements
Project performance measurement baseline
Approved change requests
List of deliverables
Other in-house (custom) factors of relevance and interest

Tools and Techniques
Information presentation tools
Performance information gathering and compilation
Status review meetings
Time reporting systems
Cost reporting systems
Other in-house (custom) tools and techniques

Output(s)
Performance reports
Forecasts
Requested changes
Recommended corrective actions
Organizational process assets
Other in-house outputs, reports, and data inferences of interest to the
  organization

time. That is, a combination of 2 from n team members. That is, the number of possible combinations of 2 members out of a team of n people. Thus, the formula for communication complexity reduces to the expression below, after some of the computation factors cancel out:

$$_nC_2 = \frac{n(n-1)}{2}$$

TABLE 7.5   Tools and Techniques for Managing Stakeholders within Project Communications Management

STEP 4: Manage Stakeholders

Inputs
Communications management plan
Organizational process assets
Other in-house (custom) factors of relevance and interest

Tools and Techniques
Communications methods
Issue logs
Other in-house (custom) tools and techniques

Output(s)
Resolved issues
Conflict resolution report
Approved change requests
Approved corrective actions
Organizational process assets (updates)
Other in-house outputs, reports, and data inferences of interest to the
  organization

In a similar vein, Badiru (2008) introduced a formula for cooperation complexity based on the statistical concept of permutation. The permutation is the number of possible arrangements of k objects taken from a set of n objects. The permutation formula is written as:

$$_nP_k = \frac{n!}{(n-k)!}$$

Thus, for the number of possible permutations of 2 members out of a team of n members is estimated as:

$$_nP_2 = n(n-1)$$

Permutation formula is used for cooperation because cooperation is bi-directional. Full cooperation requires that if A cooperates with B, then B must cooperate with A. But, A cooperating with

B does not necessarily imply B cooperating with A. In notational form, that is:

$$\mathbf{A} \rightarrow \mathrm{B} \text{ does not necessarily imply } \mathrm{B} \rightarrow \mathrm{A}.$$

Communication complexity increases rapidly as the number of communication participants increases. Coordination complexity is even more exponential as the number of team members increases. Interested readers can derive their own coordination complexity formula based on the standard combination and permutation formulas or other statistical measures. The complexity formulas indicate a need for a more structured approach to implementing the techniques of project management. The communications templates and guidelines presented in this chapter are useful for general management of STEP projects. Each specific project implementation must adapt the guidelines to the prevailing scenario and constraints of a project.

COMMUNICATING THROUGH TRIPLE C MODEL

Badiru (2008) presents the Triple C model as an effective tool for achieving communication, cooperation, and coordination in a complex project environment. The Triple C model states that project management can be enhanced by implementing it within the following integrated and hierarchical processes:

- Communication

- Cooperation

- Coordination

The model facilitates a systematic approach to project planning, organizing, scheduling, and control. The Triple C model requires communication to be the first and foremost function in the project endeavor. The model explicitly provides an avenue to address questions such as the following:

When will the project be accomplished?

Which tools are available for the project?

What training is needed for the project execution?

What resources are available for the project?

Who will be part of the project team?

Triple C model involves how the basic questions of what, who, why, how, where, and when revolve around project tasks. It highlights what must be done and when. It can also help to identify the resources (personnel, equipment, facilities, etc.) required for each effort in the project. It points out important questions such as

- Does each project participant know what the objective is?

- Does each participant know his or her role in achieving the objective?

- What obstacles may prevent a participant from playing his or her role effectively?

Triple C can mitigate the disparity between idea and practice because it explicitly solicits information about the critical aspects of a project. The different types of communication, cooperation, and coordination are outlined below.

Types of Communication
- Verbal

- Written

- Body language

- Visual tools (e.g., graphical tools)

- Sensual (Use of all five senses: sight, smell, touch, taste, hearing – olfactory, tactile, auditory)

- Simplex (unidirectional)
- Half-duplex (bi-directional with time lag)
- Full-duplex (real-time dialogue)
- One-on-one
- One-to-many
- Many-to-one

Types of Cooperation

- Proximity
- Functional
- Professional
- Social
- Romantic
- Power influence
- Authority influence
- Hierarchical
- Lateral
- Cooperation by intimidation
- Cooperation by enticement

Types of Coordination

- Teaming
- Delegation
- Supervision

- Partnership

- Token-passing

- Baton hand-off

TYPICAL TRIPLE C QUESTIONS

Questioning is the best approach to getting information for effective project management. Everything should be questioned. By upfront questions, we can preempt and avert project problems later on. Typical questions to ask under the Triple C approach are:

- What is the purpose of the project?

- Who is in charge of the project?

- Why is the project needed?

- Where is the project located?

- When will the project be carried out?

- How will the project contribute to increased opportunities for the organization?

- What is the project designed to achieve?

- How will the project affect different groups of people within the organization?

- What will the project approach or methodology be?

- What other groups or organizations will be involved (if any)?

- What will happen at the end of the project?

- How will the project be tracked, monitored, evaluated, and reported?

- What resources are required?

- What are the associated costs of the required resources?

- How do the project objectives fit the goal of the organization?

- What respective contribution is expected from each participant?

- What level of cooperation is expected from each group?

- Where is the coordinating point for the project?

TRIPLE C COMMUNICATION

Communication makes working together possible. The communication function of project management involves making all those concerned become aware of project requirements and progress. Those who will be affected by the project directly or indirectly, as direct participants or as beneficiaries, should be informed as appropriate regarding the following:

- Scope of the project

- Personnel contribution required

- Expected cost and merits of the project

- Project organization and implementation plan

- Potential adverse effects if the project should fail

- Alternatives, if any, for achieving the project goal

- Potential direct and indirect benefits of the project

The communication channel must be kept open throughout the project life cycle. In addition to internal communication, appropriate external sources should also be consulted. The project manager must

- Exude commitment to the project

- Utilize the communication responsibility matrix

- Facilitate multi-channel communication interfaces
- Identify internal and external communication needs
- Resolve organizational and communication hierarchies
- Encourage both formal and informal communication links

When clear communication is maintained between management and employees and among peers, many project problems can be averted. Project communication may be carried out in one or more of the following formats:

- One-to-many
- One-to-one
- Many-to-one
- Written and formal
- Written and informal
- Oral and formal
- Oral and informal
- Nonverbal gestures

Good communication is affected when what is implied is perceived as intended. Effective communications are vital to the success of any project. Despite the awareness that proper communications form the blueprint for project success, many organizations still fail in their communications functions. The study of communication is complex. Factors that influence the effectiveness of communication within a project organization structure include the following.

1. **Personal perception**: Each person perceives events on the basis of personal psychological, social, cultural, and experimental background. As a result, no two people can interpret a given event the same way. The nature of events is not always the critical aspect of a problem situation. Rather, the problem is often the different perceptions of the different people involved.

2. **Psychological profile**: The psychological makeup of each person determines personal reactions to events or words. Thus, individual needs and level of thinking will dictate how a message is interpreted.

3. **Social environment**: Communication problems sometimes arise because people have been conditioned by their prevailing social environment to interpret certain things in unique ways. Vocabulary, idioms, organizational status, social stereotypes, and economic situation are among the social factors that can thwart effective communication.

4. **Cultural background**: Cultural differences are among the most pervasive barriers to project communications, especially in today's multinational organizations. Language and cultural idiosyncrasies often determine how communication is approached and interpreted.

5. **Semantic and syntactic factors**: Semantic and syntactic barriers to communications usually occur in written documents. Semantic factors are those that relate to the intrinsic knowledge of the subject of the communication. Syntactic factors are those that relate to the form in which the communication is presented. The problems created by these factors become acute in situations where response, feedback, or reaction to the communication cannot be observed.

6. **Organizational structure**: Frequently, the organization structure in which a project is conducted has a direct influence on the flow of information and, consequently, on the effectiveness of communication. Organization hierarchy may determine how different personnel levels perceive a given communication.

7. **Communication media**: The method of transmitting a message may also affect the value ascribed to the message and consequently, how it is interpreted or used. The common barriers to project communications are

- Inattentiveness
- Lack of organization
- Outstanding grudges
- Preconceived notions
- Ambiguous presentation
- Emotions and sentiments
- Lack of communication feedback
- Sloppy and unprofessional presentation
- Lack of confidence in the communicator
- Lack of confidence by the communicator
- Low credibility of communicator
- Unnecessary technical jargon
- Too many people involved
- Untimely communication
- Arrogance or imposition
- Lack of focus.

Some suggestions to improve the effectiveness of communication are presented next. The recommendations may be implemented as appropriate for any of the forms of communications listed earlier. The recommendations are for both the communicator and the audience.

1. Never assume that the integrity of the information sent will be preserved as the information passes through several communication channels. Information is generally filtered, condensed, or expanded by the receivers before relaying it to the next destination. When preparing a communication that needs to pass through several organization structures, one safeguard is to compose the original information in a concise form to minimize the need for re-composition of the project structure.

2. Give the audience a central role in the discussion. A leading role can help make a person feel a part of the project effort and responsible for the projects' success. He or she can then have a more constructive view of project communication.

3. Do homework and think through the intended accomplishment of the communication. This helps eliminate trivial and inconsequential communication efforts.

4. Carefully plan the organization of the ideas embodied in the communication. Use indexing or points of reference whenever possible. Grouping ideas into related chunks of information can be particularly effective. Present the short messages first. Short messages help create focus, maintain interest and prepare the mind for the longer messages to follow.

5. Highlight why the communication is of interest and how it is intended to be used. Full attention should be given to the content of the message with regard to the prevailing project situation.

6. Elicit the support of those around you by integrating their ideas into the communication. The more people feel they have contributed to the issue, the more expeditious they are in soliciting the cooperation of others. The effect of the multiplicative rule can quickly garner support for the communication purpose.

7. Be responsive to the feelings of others. It takes two to communicate. Anticipate and appreciate the reactions of members of the audience. Recognize their operational circumstances and present your message in a form they can relate to.

8. Accept constructive criticism. Nobody is infallible. Use criticism as a springboard to higher communication performance.

9. Exhibit interest in the issue in order to arouse the interest of your audience. Avoid delivering your messages as a matter of a routine organizational requirement.

10. Obtain and furnish feedback promptly. Clarify vague points with examples.

11. Communicate at the appropriate time, in the right place, to the right people.

12. Reinforce words with positive action. Never promise what cannot be delivered. Value your credibility.

13. Maintain eye contact in oral communication and read the facial expressions of your audience to obtain real-time feedback.

14. Concentrate on listening as much as speaking. Evaluate both the implicit and explicit meanings of statements.

15. Document communication transactions for future references.

16. Avoid asking questions that can be answered yes or no. Use relevant questions to focus the attention of the audience. Use questions that make people reflect upon their words, such as, "How do you think this will work?" compared to "Do you think this will work?"

17. Avoid patronizing the audience. Respect their judgment and knowledge.

18. Speak and write in a controlled tempo. Avoid emotionally charged voice inflections.

19. Create an atmosphere for the formal and informal exchange of ideas.

20. Summarize the objectives of the communication and how they will be achieved.

SMART COMMUNICATION

The key to getting everyone on board with a project is to ensure that task objectives are clear and comply with the principle of **SMART** as outlined below:

Specific: Task objective must be specific.

Measurable: Task objective must be measurable.

Aligned: Task objective must be achievable and aligned with the overall project goal.

Realistic: Task objective must be realistic and relevant to the organization.

Timed: Task objective must have a time basis.

If a task has the above intrinsic characteristics, then the function of communicating the task will more likely lead to personnel cooperation. A communication responsibility matrix shows the linking of sources of communication and targets of communication. Cells within the matrix indicate the subject of the desired communication. There should be at least one filled cell in each row and each column of the matrix. This assures that each individual of a department has at least one communication source or target associated with him or her. With a communication responsibility matrix, a clear understanding of what needs to be communicated to whom can be developed. Communication in a project environment can take any of several forms. The specific needs of a project may dictate the most appropriate mode. Three popular computer communication modes are discussed next in the context of communicating data and information for project management.

Simplex Communication. This is a unidirectional communication arrangement in which one project entity initiates communication to another entity or individual within the project environment. The entity addressed in the communication does not have mechanism or capability for responding to the communication. An extreme example of this is a one-way, top-down communication from top management to the project personnel. In this case, the personnel have no communication access or input to top management. A budget-related example is a case where top management allocates a budget to a project without requesting and reviewing the actual needs of the project. Simplex communication is common in authoritarian organizations.

Half-Duplex Communication. This is a bi-directional communication arrangement whereby one project entity can communicate with another entity and receive a response within a certain time lag. Both entities can communicate with each other but not at the same time. An example of half-duplex communication is a project organization that permits communication with top management without a direct meeting. Each communicator must wait for a response from the target of the communication. Request and allocation without a budget meeting is another example of half-duplex data communication in project management.

Full-Duplex Communication. This involves a communication arrangement that permits a dialogue between the communicating entities. Both individuals and entities can communicate with each other at the same time or face-to-face. As long as there is no clash of words, this appears to be the most receptive communication mode. It allows participative project planning in which each project participant has an opportunity to contribute to the planning process.

Each member of a project team needs to recognize the nature of the prevailing communication mode in the project. Management must evaluate the prevailing communication structure and attempt to modify it if necessary to enhance project functions. An evaluation of who is to communicate with whom about what may help improve the project data/information communication process. A communication matrix may include notations about the desired modes of communication between individuals and groups in the project environment.

TRIPLE C COOPERATION

The cooperation of the project personnel must be explicitly elicited. Merely voicing consent for a project is not enough assurance of full cooperation. The participants and beneficiaries of the project must be convinced of the merits of the project. Some of the factors that influence cooperation in a project environment include personnel requirements, resource requirements, budget limitations, past experiences, conflicting priorities, and lack of uniform organizational support. A structured approach to seeking cooperation should clarify the following:

- Cooperative efforts required

- Precedents for future projects

- Implication of lack of cooperation

- Criticality of cooperation to project success

- Organizational impact of cooperation

- Time-frame involved in the project

- Rewards of good cooperation

Cooperation is a basic virtue of human interaction. More projects fail due to a lack of cooperation and commitment than any other project factors. To secure and retain the cooperation of project participants, you must elicit a positive first reaction to the project. The most positive aspects of a project should be the first items of project communication. For project management, there are different types of cooperation that should be understood.

Functional Cooperation. This is cooperation induced by the nature of the functional relationship between two groups. The two groups may be required to perform related functions that can only be accomplished through mutual cooperation.

Social Cooperation. This is the type of cooperation affected by the social relationship between two groups. The prevailing social relationship motivates cooperation that may be useful in getting project work done.

Legal Cooperation. Legal cooperation is the type of cooperation that is imposed through some authoritative requirement. In this case, the participants may have no choice other than to cooperate.

Administrative Cooperation. This is cooperation brought on by administrative requirements that make it imperative that two groups work together on a common goal.

Associative Cooperation. This type of cooperation may also be referred to as collegiality. The level of cooperation is determined by the association that exists between two groups.

Proximity Cooperation. Cooperation due to the fact that two groups are geographically close is referred to as proximity cooperation. Being close makes it imperative that the two groups work together.

Dependency Cooperation. This is cooperation caused by the fact that one group depends on another group for some important aspect. Such dependency is usually of a mutual two-way nature. One group depends on the other for one thing while the latter group depends on the former for some other thing.

Imposed Cooperation. In this type of cooperation, external agents must be employed to induce cooperation between two groups. This is applicable for cases where the two groups have no natural reason to cooperate. This is where the approaches presented earlier for seeking cooperation can become very useful.

Lateral Cooperation. Lateral cooperation involves cooperation with peers and immediate associates. Lateral cooperation is often easy to achieve because existing lateral relationships create an environment that is conducive for project cooperation.

Vertical Cooperation. Vertical or hierarchical cooperation refers to cooperation that is implied by the hierarchical structure of the project. For example, subordinates are expected to cooperate with their vertical superiors.

Whichever type of cooperation is available in a project environment; the cooperative forces should be channeled toward achieving project goals. Documentation of the prevailing level of cooperation is useful for winning further support for a project. Clarification of project priorities will facilitate personnel cooperation. Relative priorities of multiple projects should be specified so that they are a priority to all groups within the organization. Some guidelines for securing cooperation for most projects are

- Establish achievable goals for the project

- Clearly outline the individual commitments required

- Integrate project priorities with existing priorities

- Eliminate the fear of job loss due to industrialization

- Anticipate and eliminate potential sources of conflict

- Use an open-door policy to address project grievances

- Remove skepticism by documenting the merits of the project

Commitment. Cooperation must be supported with commitment. To cooperate is to support the ideas of a project. To commit is to willingly and actively participate in project efforts again and again through the thick and thin of the project. Provision of resources is one way that management can express commitment to a project.

TRIPLE C COORDINATION

After the communication and cooperation functions have successfully been initiated, the efforts of the project personnel must be coordinated. Coordination facilitates harmonious organization of project efforts. The construction of a responsibility chart can be very helpful at this stage. A responsibility chart is a matrix consisting of columns of individual or functional departments and rows of required actions. Cells within the matrix are filled with relationship codes that indicate who is responsible for what.

- Who is to do what?

- How long will it take?

- Who is to inform whom of what?

- Whose approval is needed for what?

- Who is responsible for which results?

- What personnel interfaces are required?

- What support is needed from whom and when?

CONFLICT RESOLUTION USING TRIPLE C APPROACH

Conflicts can and do develop in any work environment. Conflict, whether intended or inadvertent, prevents an organization from getting the most out of the workforce. When implemented as an integrated process, the Triple C model can help avoid conflicts in a project. When conflicts do develop, it can help in resolving the conflicts. The key to conflict resolution is open and direct communication, mutual cooperation, and sustainable coordination. Several sources of conflicts can exist in a project. Some of these are discussed below.

Schedule Conflict. Conflicts can develop because of improper timing or sequencing of project tasks. This is particularly common in large multiple projects. Procrastination can lead to having too much to do at once, thereby creating a clash of project functions and discord among project team members. Inaccurate estimates of time requirements may lead to infeasible activity schedules. Project coordination can help avoid schedule conflicts.

Cost Conflict. Project cost may not be generally acceptable to the clients of a project. This will lead to project conflict. Even if the initial cost of the project is acceptable, a lack of cost control during implementation can lead to conflicts. Poor budget allocation approaches and the lack of a financial feasibility study will cause cost conflicts later on in a project. Communication and coordination can help prevent most of the adverse effects of cost conflicts.

Performance Conflict. If clear performance requirements are not established, performance conflicts will develop. Lack of clearly defined performance standards can lead each person to evaluate his or her own performance based on personal value judgments. In order to uniformly evaluate the quality of the work and monitor project progress, performance standards should be established by using the Triple C approach.

Management Conflict. There must be a two-way alliance between management and the project team. The views of management should be understood by the team. The views of the team should be appreciated by management. If this does not happen, management conflicts will develop. A lack of a two-way interaction can lead to strikes and industrial actions, which can be detrimental to project objectives. The Triple C approach can help create a conducive dialogue environment between management and the project team.

Technical Conflict. If the technical basis of a project is not sound, technical conflict will develop. New industrial projects are particularly prone to technical conflicts because of their significant dependence on technology. A lack of a comprehensive technical feasibility study will lead to technical conflicts. Performance requirements and systems specifications can be integrated through the Triple C approach to avoid technical conflicts.

Priority Conflict. Priority conflicts can develop if project objectives are not defined properly and applied uniformly across a project. A lack of a direct project definition can lead each project member to define his or her own goals which may be in conflict with the intended goal of a project. A lack of consistency of the project mission is another potential source of conflict in priorities. Over-assignment of responsibilities with no guidelines for relative significance levels can also lead to priority conflicts. Communication can help defuse priority conflict.

Resource Conflict. Resource allocation problems are a major source of conflict in project management. Competition for resources, including personnel, tools, hardware, software, and so on, can lead to disruptive clashes among project members. The Triple C approach can help secure resource cooperation.

Power Conflict. Project politics lead to power plays which can adversely affect the progress of a project. Project authority and project power should be clearly delineated. Project authority is the control that a person has by virtue of his or her functional post.

Project power relates to the clout and influence, which a person can exercise due to connections within the administrative structure. People with popular personalities can often wield a lot of project power in spite of low or nonexistent project authority. The Triple C model can facilitate a positive marriage of project authority and power to the benefit of project goals. This will help define clear leadership for a project.

Personality Conflict. Personality conflict is a common problem in projects involving a large group of people. The larger the project, the larger the size of the management team needed to keep things running. Unfortunately, the larger management team creates an opportunity for personality conflicts. Communication and cooperation can help defuse personality conflicts. In summary, conflict resolution through Triple C can be achieved by observing the following guidelines:

1. Confront the conflict and identify the underlying causes.

2. Be cooperative and receptive to negotiation as a mechanism for resolving conflicts.

3. Distinguish between proactive, inactive, and reactive behaviors in a conflict situation.

4. Use communication to defuse internal strife and competition.

5. Recognize that short-term compromise can lead to long-term gains.

6. Use coordination to work toward a unified goal.

7. Use communication and cooperation to turn a competitor into a collaborator.

It is the little and often neglected aspects of a project that lead to project failures. Several factors may constrain the project implementation. All the relevant factors can be evaluated under the Triple C model right from the project-initiation stage.

BIBLIOGRAPHY

Badiru, A. B., *Triple C Model of Project Management*, Taylor & Francis Group, CRC Press, London, UK, Boca Raton, FL, 2008.

Mooz, H.; K. Forsberg; and H. Cotterman, *Communicating Project Management*, John Wiley & Sons, Hoboken, New Jersey, 2003.

PMI, *A Guide to the Project Management Body of Knowledge (PMBOK Guide)*, 6th Edition, Project Management Institute, Philadelphia, PA, 2017.

CHAPTER 8

Risk Management

P ROJECT MANAGEMENT IS ABOUT taking risk, venturing out and discovering what is out there, and exploring what exists within or outside the realm of possibility. Risk management is the process of identifying, analyzing, and recognizing the various risks and uncertainties that might affect a project. Change can be expected in any project environment. Change portends risk and uncertainty. Risk analysis outlines possible future events and their likelihood of occurrence. With the information from risk analysis, the project team can be better prepared for change with good planning and control actions. By identifying the various project alternatives and their associated risk, the project team can select the most appropriate courses of action.

Risk permeates every aspect of a project. In fact, each and every one of the other elements in the project management body of knowledge is subject to some level of risk. Scope presents risks. Communication has risk components. Cost is subject to risk. Time has factors of risk and uncertainty. Quality variability contains a dimension of risk. Human resources pose operational risks. Procurement is subject to risk realities of the marketplace. Just as risk presents opportunities, it also poses threats. Thus, risk management is a crucial component of project management, particularly in science and technology-based projects where systems dynamics can disrupt operations in a flash.

RISK DEFINITION

PMI's PMBOK defines risk as "an uncertain event or condition that, if it occurs, has a positive or negative effect on at least one project objective, such as time, cost, scope, or quality." In risk management, it is assumed that a number of possible future states of a variable exist. Each occurrence of the variable has a known or assumed probability of occurring. There are often interdependencies in factors associated with a risk event. Thus, quantitative assessment is often very complex. Once a risk occurs, it is no longer a risk; it is a fact. There are three elements of risk:

1. There is some future event that has not occurred yet.
2. There is some level of uncertainty associated with the event.
3. There is a consequence (positive or negative) emanating from the risk event.

Risk management is the process of identifying, analyzing, and recognizing the various risks and uncertainties that might affect a project. The purpose of risk management is to achieve one of the following:

- Maximize the probability and consequence of positive events
- Minimize the probability and consequence of negative events

There are three possible risk response behaviors for risk management:

1. Risk-averse behavior: conscious and deliberate attempt to avoid risk

2. Risk-seeking behavior: conscious and deliberate pursuit of risk, perhaps as a manifestation of the old West saying that "you cannot accumulate if you don't speculate"

3. Risk-neutral behavior:- indifference to the presence or absence of risk

The typical recommended level of investment in risk management is around 5% to 10% of the total project budget. If there is no risk management plan in a project, then the project is operating in a fire-fighting mode. This is an example of management by exception (MBE), under which an organization does not make contingency plans.

RISK MANAGEMENT: STEP-BY-STEP IMPLEMENTATION

The risk management component of the project management body of knowledge consists of the elements shown below. The six elements are carried out across the process groups. The overlay of the elements and the process groups are shown in Table 8.1. Thus, under the knowledge area of communications management, the required steps are:

Step 1: Risk Management Planning

Step 2: Risk Identification

TABLE 8.1   Implementation of Project Risk Management Across Process Groups

Initiating
Planning
1. Risk Management Planning
2. Risk Identification
3. Qualitative Risk Analysis
4. Quantitative Risk Analysis
5. Risk Response Planning
Executing
Monitoring and controlling
1. Risk monitoring and control
Closing

Step 3: Qualitative Risk Analysis

Step 4: Quantitative Risk Analysis

Step 5: Risk Response Planning

Step 6: Risk Monitoring And Control

Tables 8.2 through 8.7 present the inputs, tools, techniques, and outputs of each step of risk management. It should be emphasized that risk itself is not identified in the risk management planning phase. The planning phase is used only to identify the processes that will be used to handle risk. Also, risk identification simply develops a list of risks; it does not rank or analyze the risks. Risk register, which is an output of risk identification, presents a list (preferably a spreadsheet) of risk events, their root causes, and associated responses. Risk rating matrix, which is a tool for qualitative risk analysis, presents a matrix of risk events with respect to their respective probabilities of occurrence and impact levels. The probability ranges are presented as:

Near Certainty, Highly Likely, Likely, Unlikely, Remote.

TABLE 8.2   Tools and Techniques for Risk Planning within Project Risk Management

STEP 1: Risk Planning
Inputs
Enterprise environmental factors
Organizational process assets
Project scope statement
Other in-house (custom) factors of relevance and interest
Tools and Techniques
Communications requirement analysis
Planning meetings and analysis
Other in-house (custom) tools and techniques
Output(s)
Risk management plan
Other in-house outputs, reports, and data inferences of interest to the organization

TABLE 8.3   Tools and Techniques for Risk Identification within Project Risk
Management

STEP 2: Risk Identification

Inputs
Enterprise environmental factors
Organizational process assets
Project scope statement
Risk management plan
Project management plan
Other in-house (custom) factors of relevance and interest

Tools and Techniques
Documentation reviews
Information-gathering techniques
Survey of subject matter experts to get risk information
Checklist analysis
Assumptions analysis
Diagramming techniques
Other in-house (custom) tools and techniques

Output(s)
Risk register
Other in-house outputs, reports, and data inferences of interest to the
    organization

The impact levels are presented as:

Negligible, Minor, Moderate, Serious, and Critical.

PROJECT DECISIONS UNDER RISK AND UNCERTAINTY

Traditional decision theory classifies decisions under three different influences:

- **Decision under certainty**: Made when possible event(s) or
    outcome(s) of a decision can be positively determined

- **Decisions under risk**: Made using information on the prob-
    ability that a possible event or outcome will occur

- **Decisions under uncertainty**: Made by evaluating possible
    event(s) or outcome(s) without information on the probabil-
    ity that the event(s) or outcome(s) will occur

TABLE 8.4   Tools and Techniques for Qualitative Risk Analysis within Project Risk Management

STEP 3: Qualitative Risk Analysis
Inputs
Organizational process assets
Project scope statement
Risk management plan
Risk register
Other in-house (custom) factors of relevance and interest
Tools and Techniques
Critical incident safety management (CISM) for risk analysis
Risk rating matrix
Risk probability and impact assessment
Probability and impact matrix
Risk data quality assessment
Risk categorization
Risk urgency assessment
Other in-house (custom) tools and techniques
Output(s)
Risk register (updates)
Other in-house outputs, reports, and data inferences of interest to the organization

Many authors make a distinction between decisions under risk and under uncertainty. In the literature, decisions made under uncertainty increasingly incorporate decisions made under risk, as defined previously. In this book, no special distinction will be made between risk and uncertainty. Some of the chapters in this book contain a number of procedures to illustrate how project decisions may be made under uncertainty. Some of the parameters that normally change during a project lifecycle include project costs, time requirements, and performance specifications. The uncertainties associated with these parameters are a concern for project managers. Cost, time, and performance must be managed throughout the project lifecycle.

COST UNCERTAINTIES

In an inflationary economy, project costs can become very dynamic and intractable. Cost estimates include various tangible

TABLE 8.5   Tools and Techniques for Quantitative Risk Analysis within
Project Risk Management

STEP 4: Quantitative Risk Analysis
Inputs
Organizational process assets
Project scope statement
Risk management plan
Risk register
Project management plan
Other in-house (custom) factors of relevance and interest
Tools and Techniques
Data gathering and representation techniques
Quantitative risk analysis and modeling techniques
Other in-house (custom) tools and techniques
Output(s)
Risk register (updates)
Other in-house outputs, reports, and data inferences of interest to the organization

and intangible components of a project, such as machines, inventory, training, raw materials, design, and personnel wages. Costs can change during a project for a number of reasons including:

- External inflationary trends

- Internal cost adjustment procedures

- Modification of work process

- Design adjustments

- Changes in cost of raw materials

- Changes in labor costs

- Adjustment of work breakdown structure

- Cash flow limitations

- Effects of tax obligations

TABLE 8.6   Tools and Techniques for Risk Response Planning within Project Risk Management

STEP 5: Risk Response Planning
Inputs
Risk management plan
Risk register
Other in-house (custom) factors of relevance and interest
Tools and Techniques
Risk tolerance level
Risk-averse tendencies
Strategies for negative risks or threats
Strategies for positive risks or opportunities
Contingency strategy
Other in-house (custom) tools and techniques
Output(s)
Risk register (updates)
Project management plan (updates)
Risk-related contractual agreements
Other in-house outputs, reports, and data inferences of interest to the organization

These cost changes and others combine to create uncertainties in the project's cost. Even when the cost of some of the parameters can be accurately estimated, the overall project cost may still be uncertain due to the few parameters that cannot be accurately estimated.

SCHEDULE UNCERTAINTIES

Unexpected engineering change orders (ECO) and other changes in a project environment may necessitate schedule changes, which introduce uncertainties to the project. The following are some of the reasons project schedules change:

- Task adjustments
- Changes in scope of work
- Changes in delivery arrangements

TABLE 8.7   Tools and Techniques for Risk Monitoring and Control within
Project Risk Management

STEP 6: Risk Monitoring and Control

Inputs
Risk management plan
Risk register
Approved change requests
Work performance information
Performance reports
Other in-house (custom) factors of relevance and interest

Tools and Techniques
Risk reassessment
Risk mitigation method
Risk audits
Variance and trend analysis
Technical performance measurement
Reserve analysis
Status meetings
Other in-house (custom) tools and techniques

Output(s)
Risk register (updates)
Requested changes
Recommended preventive actions
Organizational process assets (updates)
Project management plan (updates)
Other in-house outputs, reports, and data inferences of interest to the organization

- Changes in the project specification

- Introduction of new technology

PERFORMANCE UNCERTAINTIES

Performance measurement involves observing the value of the
parameter(s) during a project, and comparing the actual performance, based on the observed parameter(s), to the expected
performance. Performance control then takes appropriate actions
to minimize the deviations between actual performance and

expected performance. Project plans are based on the expected performance of the project parameters. Performance uncertainties exist when expected performance cannot be defined in definite terms. As a result, project plans require a frequent review.

The project management team must have a good understanding of the factors that can have a negative impact on the expected project performance. If at least some of the sources of deficient performance can be controlled, then the detrimental effects of uncertainties can be alleviated. The most common factors that can influence project performance include the following:

- Redefinition of project priorities

- Changes in management control

- Changes in resource availability

- Changes in work ethic

- Changes in organizational policies and procedures

- Changes in personnel productivity

- Changes in quality standards

To minimize the effect of uncertainties in project management, a good control must be maintained over the various sources of uncertainty discussed above. The same analytic tools that are effective for one category of uncertainties should also work for other categories.

RISK AND DECISION TREES

Decision tree analysis is used to evaluate sequential decision problems. In project management, a *decision tree* may be useful for evaluating sequential project milestones. A decision problem under certainty has two elements: *action and consequence*. The decision maker's choices are the actions while the results of those actions are the consequences. For example, in a CPM network planning, the choice of one task among three potential tasks in a given time slot represents a potential action. The consequences of choosing one task over another may be characterized in terms of the slack time created in the network, the cost of performing the selected task, the resulting effect on the project completion time, or the degree to which a specified performance criterion is satisfied.

If the decision is made under uncertainty, as in PERT network analysis, a third element, called an *event*, is introduced into the decision problem. Extending the CPM task selection example to a PERT analysis, the actions may be defined as Select Task 1, Select Task 2, and Select Task 3. The durations associated with the three possible actions can be categorized as *long task duration, medium task duration,* and *short task duration.* The actual duration of each task is uncertain. Thus, each task has some probability of exhibiting long, medium, or short durations.

The events can be identified as weather incidents: rain or no rain. The incidents of rain or no rain are uncertain. The consequences may be defined as *increased project completion time, decreased project completion time,* and *unchanged project completion time.* These consequences are also uncertain due to the probable durations of the tasks and the variable choices of the decision maker. That is, the consequences are determined partly by choice and partly by chance. The consequences also depend on which event occurs—rain or no rain.

To simplify the decision analysis, the decision elements may be summarized by using a decision table. A *decision table* shows the relationship between pairs of decision elements.

In some decision problems, the consequences may not be unique. Thus, a consequence, which is associated with a particular event–action pair, may also be associated with another event–action pair. The actions included in the decision table are the only ones that the decision maker wishes to consider. Subcontracting and task elimination, for example, are other possible choices for the decision maker. The actions included in the decision problem are mutually exclusive and collectively exhaustive, so that exactly one will be selected. The events are also mutually exclusive and collectively exhaustive.

CHAPTER 9

Procurement
Management

PROCUREMENT MANAGEMENT INVOLVES THE process of acquiring the necessary equipment, tools, goods, services, and resources needed to successfully accomplish project goals. Procurement is often called acquisition, purchasing, or contracting. This represents the process of acquiring (through contracting) products, results, or services for direct usage on a project (Badiru, 1995; Badiru, 2007; Chopra and Meindl, 2004; PMI, 2017). The end results of a project fall in three major categories of:

- Products

- Services

- Results

Procurement is needed as a formal process of obtaining the above from a vendor or supplier whether the products or services are already in existence or must be newly designed, developed, tested, or demonstrated. Procurement involves all aspects of contract administration during the project lifecycle. The buy, lease, or make options available to the project must be evaluated with respect to time, cost, and technical performance requirements. Contractual agreements, in written or unwritten (verbal) format, constitute the legal document that defines the work obligation of each participant in a project. Procurement refers to the actual process of obtaining the needed services and resources. A contract, within the context of project procurement, is a mutually binding agreement that obligates the vendor to provide the specified products, services, or results and obligates the buyer to provide a monetary return for the contract rendered. The procurement cycle occurs at the project–supplier interface and covers all processes necessary to ensure that materials are available for executing the project schedule. The supply chain networking becomes very essential during the procurement cycle.

Coordinated procurement is particularly crucial for science, technology, and engineering projects. Sourcing, within the procurement process, involves the selection of suppliers, development of contracts, product design collaboration, materials supply, and evaluation of vendor performance. Just like any partnership relationship, the project management team must cultivate, nurture, and sustain a positive alliance with vendors for the project; and the alliance must center around the following dimensions of partnership:

- Project-vendor communication

- Project-vendor cooperation

- Project-vendor coordination

PROCUREMENT MANAGEMENT: STEP-BY-STEP IMPLEMENTATION

The procurement management component of the project management body of knowledge consists of the elements shown below. The six elements are carried out across the process groups. The

TABLE 9.1   Implementation of Project Procurement Management Across
Process Groups

Initiating
Planning
1. Plan Purchases and Acquisitions
2. Plan Contracting
Executing
1. Request Seller Responses
2. Select Sellers
Monitoring and Controlling
1. Contract Administration
Closing

overlay of the elements and the process groups are shown in Table 9.1. Thus, under the knowledge area of communications management, the required steps are:

Step 1: Plan Purchases and Acquisitions

Step 2: Plan Contracting

Step 3: Request Seller Responses

Step 4: Select Vendors (Sellers)

Step 5: Contract Administration

Step 6: Contract Closure

Tables 9.2 through 9.7 present the inputs, tools, techniques, and outputs of each step in procurement management. Plan purchases and acquisitions constitutes the process of identifying which components of a project to acquire through the procurement process. This involves the following queries:

- Whether or not to acquire the component

- How to acquire the component

- What to acquire

TABLE 9.2   Tools and Techniques for Purchases and Acquisitions within
Project Procurement Management

STEP 1: Plan Purchases and Acquisitions

Inputs
Enterprise environmental factors
Organizational process assets
Project scope statement
WBS dictionary
Project management plan
Other in-house (custom) factors of relevance and interest

Tools and Techniques
Make-or-buy analysis
Breakeven analysis
Expert judgment
Contract type selection
Project selection criteria
Minimum revenue requirement analysis
Other in-house (custom) tools and techniques

Output(s)
Procurement management plan
Contractor statement of work
Make-or-buy decision
Requested changes
Other in-house outputs, reports, and data inferences of interest to the
  organization

- How much of the component to acquire

- When to acquire the component

Explanations of the entries in the step-by-step tables are provided below:

Enterprise environmental factors describe marketplace conditions, what is available, from whom, and in what quantity and quality.

Organizational process assets provide the formal and informal policies, procedures, guidelines, and management systems for the procurement management plan and contract type.

TABLE 9.3   Tools and Techniques for Contracting within Project Procurement Management

STEP 2: Plan Contracting

Inputs
Procurement management plan
Contract statement of work (CSOW)
Make-or-Buy Decisions
Project management plan
Other in-house (custom) factors of relevance and interest

Tools and Techniques
Make-or-buy analysis
Breakeven analysis
Contracting standard forms
Contract administration planning
Expert judgment
Other in-house (custom) tools and techniques

Output(s)
Procurement documents
Evaluation criteria
CSOW updates
Other in-house outputs, reports, and data inferences of interest to the
   organization

Project scope statement describes project boundaries, requirements, (e.g., safety clearance and permit), constraints (e.g., budget limitation), and assumptions (e.g., resource availability) related to the project scope.

Work breakdown structure (WBS) provides the relationship among project components and deliverables.

WBS Dictionary identifies the deliverable with a description of work for each WBS component.

Project management plan provides an overall plan and includes the procurement management plan; including other considerations such as risk register for risks, owners, and risk responses, risk-related contractual agreements, insurance, activity-resource requirements, project schedule, activity cost estimates, and cost baseline.

TABLE 9.4   Tools and Techniques for Requesting Vendors within Project
Procurement Management

STEP 3: Request Vendors

Inputs
Procurement management plan
Organizational process assets
Procurement documents
Other in-house (custom) factors of relevance and interest

Tools and Techniques
Bidder conferences
Advertising
Broad-agency announcement
Development of qualified sellers list
Request for proposals (FRP)
Request for bids (RFB)
Invitation for bid (IFB)
Request for quotation (FRQ)
Invitation for negotiation (IFN)
Other in-house (custom) tools and techniques

Output(s)
Qualified sellers list
Procurement document package
Proposals
Other in-house outputs, reports, and data inferences of interest to the
  organization

Make-or-buy analysis is a general decision technique to determine whether a particular product or service can be produced
more cost-effectively organically by the project organization or
purchased from an external source. The make-or-buy analysis
reflects the interests and strategy of the project organization, the
capability of the vendor organization, as well as the immediate
needs of the project.

Expert judgment, in the context of procurement management,
assesses the inputs and outputs needed for an effective procurement
decision. Inputs would normally include the interests and oversight of other units within the project organization; including such

TABLE 9.5   Tools and Techniques for Selecting Vendors within Project Procurement Management

STEP 4: Select Vendors

Inputs
Procurement management plan
Organizational process assets
Procurement document package
Evaluation criteria
Proposals
Qualified sellers list
Project management plan
Other in-house (custom) factors of relevance and interest

Tools and Techniques
Weighting system
Independent estimates
Screening system
Contract negotiation
Vendor rating system
Expert judgment
Proposal evaluation techniques
Multi-criteria outsourcing techniques
Other in-house (custom) tools and techniques

Output(s)
Selected vendors
Contract issuance
Contract management plan
Resource availability
Update procurement management plan
Requested changes
Other in-house outputs, reports, and data inferences of interest to the
   organization

departments as legal, contracts, technical support, subject matter experts, and management preferences. In addition, inputs from external sources such as consultants, regulatory requirements, professional organizations, technical associations, and industry groups are often instrumental in making procurement decisions.

TABLE 9.6   Tools and Techniques for Contract Administration within Project Procurement Management

STEP 5: Contract Administration
Inputs
Contract
Contract management plan
Performance reports
Approved change requests
Work performance information
Other in-house (custom) factors of relevance and interest
Tools and Techniques
Contract change control system
Buyer conducted performance review
Inspection and audits
Performance reporting
Payment system
Claims administration
Records management system
Information technology
Other in-house (custom) tools and techniques
Output(s)
Contract documentation
Requested changes
Recommended corrective actions
Organization process assets (updates)
Project management plan (updates)
Other in-house outputs, reports, and data inferences of interest to the organization

Contract type selection helps to align procurement decisions with decision factors and project constraints such as cost, schedule, and performance expectations. The type of contract selected is based on the following:

- Overall cost

- Schedule compatibility

- Quality acceptance

TABLE 9.7   Tools and Techniques for Contract Closure within Project
Procurement Management

STEP 6: Contract Closure
Inputs
Procurement management plan
Contract management plan
Contract documentation
Contract closure procedure
Other in-house (custom) factors of relevance and interest
Tools and Techniques
Procurement audits
Records management systems
Other in-house (custom) tools and techniques
Output(s)
Closed contracts
Lessons learned documentation
Dissemination of project results
Organizational process assets
Other in-house outputs, reports, and data inferences of interest to the organization

- Degree of risk

- Product or service complexity (e.g., technical risk)

- Contractor's accountability, responsibility, and risk

- Concurrent contracts

- Outsourcing and subcontracting preferences

- Vendor's accounting system and reliability

- Urgency of need

Contracts fall into one of three major categories as explained
below:

1. **Fixed price or lump sum contracts** have the following
   characteristics:

    a. Fixed total price for a well-defined product or service.

    b. If the product is not well-defined, both the project and vendor are at risk.

    c. The simplest form of this is to use purchase order for a specified item, at a specified price, for a specific date.

    d. Fixed price contracts may also include incentives for meeting project objectives.

2. **Time and materials contracts** have the following characteristics:

    a. This contains aspects of both cost reimbursable and fixed price contracts.

    b. It is often open-ended and full value is usually not defined at the time of award.

    c. Unit rates for this type of contract can be preset.

3. **Cost reimbursable contracts** have the following characteristics:

    a. This involves payment to the vendor for actual costs of product or service rendered.

    b. Costs are classified as direct costs or indirect costs. Direct costs are costs incurred exclusively for the purpose of the project. Indirect costs are overhead costs that are allocated to the project by the performing organization.

Cost reimbursable contracts are further categorized into the following types:

- Cost Plus Fee (CPF) or Cost Plus Percentage of Cost (CPPC): In this case, the vendor is reimbursed for all allowable costs plus an agreed fee at an agreed percentage of costs. The fee varies with actual costs.

- Cost Plus Fixed Fee (CPFF): In this case, the vendor is reimbursed for all allowable costs plus a fixed fee payment based on a percentage of the estimated project costs. The fee does not vary with actual costs.

- Cost Plus Incentive Fee (CPIF): In this case, the vendor is reimbursed for all allowable costs and a predetermined fee (incentive bonus) based on achieving certain performance objectives. Both the vendor and buyer could benefit from cost savings on the basis of a negotiated cost formula.

COMPLETION AND TERM CONTRACTS

A contract can be executed either as a *completion contract* or a *term contract*. In a completion contract, the contractor is required to deliver a definitive end product. The contract is complete upon delivery and formal customer acceptance. The final payment is made upon delivery. In a term contract, the contractor is required to deliver a specific "level of effort," where the effort is expressed in "person-days" over a specified period of time. The contractor is under no further obligation after the effort is performed. Final payment is not dependent upon technical accomplishment.

PROCUREMENT MANAGEMENT PLAN

A procurement management plan specifies how the remaining procurement processes will be managed. It may be formal or informal, highly detailed or broadly stated, based on the specific needs of the project. It is a subsidiary of the overall project plan.

CONTRACTOR STATEMENT OF WORK

The statement of work (SOW) describes the procurement item in sufficient detail to allow prospective vendors to determine if they are capable of providing the product or service. Each individual procurement item requires a separate statement of work. However, the multiple products or services may be grouped as one procurement item with a single statement of work. Statement of work will often influence the development of additional contract evaluation criteria such as the following queries:

- Does the vendor demonstrate an understanding of the needs of the project? This can be evident in the contents of the proposal.

- What level of overall or lifecycle cost is offered by the vendor? Will the selected vendor produce the lowest total cost; including contract cost as well as operating cost?

- Does the vendor have adequate technical capability? Does the vendor currently have, or can be expected to acquire, the technical capabilities and knowledge needed by the project?

- Will the vendor's management approach ensure a successful execution of the project?

- Does the vendor have the financial status and capability adequate to execute the contract successfully and adequately?

- What certifications are available on the vendor's history, resources, and quality records?

ORGANIZATION PROCESS ASSETS

Organization process assets include historical lists of qualified vendors, past experience, and previous relationships. The list of preferred vendors is developed through some sort of rigorous methodology. Some quantitative methodologies are presented in this chapter. Bidder conferences, contractor conferences, vendor conferences, and pre-bid conferences are examples of meetings with prospective vendors prior to preparation of a proposal. The prospective vendors must have a clear understanding of the procurement process.

CONTRACT FEASIBILITY ANALYSIS

Procurement should be preceded by a formal feasibility analysis. The feasibility of a project can be ascertained in terms of technical factors, economic factors, or both. Some of the topics to be evaluated include contract responsibilities and authorities, applicable terms and laws, technical and business management approaches, and financing sources. A complex procurement process may require an independent or external negotiation process. A feasibility study is documented with a report showing all the ramifications of the project and should be broken down into the following categories:

Technical feasibility. Technical feasibility refers to the ability of the process to take advantage of the current state of the technology in pursuing further improvement. The technical capability of the personnel as well as the capability of the available technology should be considered.

Managerial feasibility. Managerial feasibility involves the capability of the infrastructure of a process to achieve and sustain process improvement. Management support, employee involvement, and commitment are key elements required to ascertain managerial feasibility.

Economic feasibility. This involves the ability of the proposed project to generate economic benefits. A benefit-cost analysis and a breakeven analysis are important aspects of evaluating the economic feasibility of new science and technology projects. The tangible and intangible aspects of a project should be translated into economic terms to facilitate a consistent basis for evaluation.

Financial feasibility. Financial feasibility should be distinguished from economic feasibility. Financial feasibility involves the capability of the project organization to raise the appropriate funds needed to implement the proposed project. Project financing can be a major obstacle in large multi-party projects because of the level of capital required. Loan availability, creditworthiness, equity, and loan schedule are important aspects of financial feasibility analysis.

Cultural feasibility. Cultural feasibility deals with the compatibility of the proposed project with the cultural setup of the project environment. In labor-intensive projects, planned functions must be integrated with the local cultural practices and beliefs. For example, religious beliefs may influence what an individual is willing to do or not do.

Social feasibility. Social feasibility addresses the influences that a proposed project may have on the social system in the project environment. The ambient social structure may be such that certain categories of workers may be in short supply or nonexistent. The effect of the project on the social status of the project participants must be assessed to ensure compatibility. It should be recognized that workers in certain industries may have certain status symbols within the society.

Safety feasibility. Safety feasibility is another important aspect that should be considered in project planning. Safety feasibility refers to an analysis of whether the project is capable of being implemented and operated safely with minimal adverse effects on the environment. Unfortunately, environmental impact assessment is often not adequately addressed in complex projects. As an example, the North America Free Trade Agreement (NAFTA) between the U.S., Canada, and Mexico was temporarily suspended in 1993 because of the legal consideration of the potential environmental impacts of the projects to be undertaken under the agreement.

Political feasibility. A politically feasible project may be referred to as a "politically correct project." Political considerations often dictate the direction for a proposed project. This is particularly true for large projects with national visibility that may have significant government inputs and political implications. For example, political necessity may be a source of support for a project regardless of the project's merits. On the other hand, worthy projects may face insurmountable opposition simply because of political factors. A political feasibility analysis requires an evaluation of the compatibility of project goals with the prevailing goals of the political system. In general, feasibility analysis for a project should include the following items:

1. **Need analysis**: This indicates recognition of a need for the project. The need may affect the organization itself, another organization, the public, or the government. A preliminary study is conducted to confirm and evaluate the need. A proposal of how the need may be satisfied is then made. Pertinent questions that should be asked include the following:

   - Is the need significant enough to justify the proposed project?

   - Will the need still exists by the time the project is completed?

   - What are alternate means of satisfying the need?

   - What are the economic, social, environmental, and political impacts of the need?

2. **Process work**: This is the preliminary analysis done to determine what will be required to satisfy the need. The work may be performed by a consultant who is an expert in the project field. The preliminary study often involves system models or prototypes. For technology-oriented projects, artist conceptions and scaled-down models may be used for illustrating the general characteristics of a process. A simulation of the proposed system can be carried out to predict the outcome before the actual project starts.

3. **Engineering and design**: This involves a detailed technical study of the proposed project. Written quotations are obtained from suppliers and subcontractors as needed. Technology capabilities are evaluated as needed. Product design, if needed, should be done at this stage.

4. **Cost estimate**: This involves estimating project cost to an acceptable level of accuracy. Levels of around minus five percent to plus fifteen percent are common at this level of a

project plan. Both the initial and operating costs are included in the cost estimation. Estimates of capital investment, recurring, and nonrecurring costs should also be contained in the cost-estimate document. Sensitivity analysis can be carried out on the estimated cost values to see how sensitive the project plan is to changes in the project scenario.

5. **Financial analysis**: This involves an analysis of the cash-flow profile of the project. The analysis should consider rates of return, inflation, sources of capital, payback periods, break-even point, residual values, and sensitivity.

6. **Project impacts**: This portion of the feasibility study provides an assessment of the impact of the proposed project. Environmental, social, cultural, political, and economic impacts may be some of the factors that will determine how a project is perceived by the public. The value-added potential of the project should also be assessed.

7. **Conclusions and recommendations**: The feasibility study should end with the overall outcome of the project analysis. This may constitute either an endorsement or disapproval of the project.

CONTENTS OF PROJECT PROPOSAL

The project proposal should present a detailed plan for executing the proposed project. The proposal may be directed to a management team within the same organization or to an external organization. The proposal contents may be written in two parts: a Technical Section and a Management Section.

TECHNICAL SECTION OF PROJECT PROPOSAL

Project background

1. Organization's expertise in the project area

2. Project scope

3. Primary objectives

4. Secondary objectives

Technical approach

- Required technology
- Available technology
- Problems and their resolutions
- Work breakdown structure

Work statement

- Task definitions and list
- Expectations

Schedule

- Gantt charts
- Milestones
- Deadlines

Project deliverables
The value of the project

- Significance
- Benefit
- Impact

MANAGEMENT SECTION OF PROJECT PROPOSAL

Project staff and experience

- Personnel credentials

Organization

- Task assignment
- Project manager, liaison, assistants, consultants, etc.

Cost analysis

- Personnel cost
- Equipment and materials
- Computing cost
- Travel
- Documentation preparation
- Cost sharing
- Facilities cost

Delivery dates

- Specified deliverables

Quality control measures

- Rework policy

Progress and performance monitoring

- Productivity measurement

Cost-control measures

- Milestone analysis
- Cost benchmarks

A contract awarded following a successful feasibility analysis conveys a legal relationship subject to remedy through the legal system. The contract will spell out the statement of work, period of performance, pricing, product support, limitation of liability, incentives, insurance, subcontractor approval, termination, and disputes resolution strategy. Requested changes from selected vendors are incorporated into the contract through the integrated change control process and constitute a part of the overall project and procurement plans.

CONTRACT TEAMWORK AND COOPERATION

We can usually get a lot done when a cohesive team exists. Using the methodology of Triple C, we can improve communication, cooperation, and coordination to manage relationships between vendors and the project team. Managing contract relationships has the following attributes:

- Ensures that the vendor's performance meets contractual requirements

- Manages the interfaces among the various providers

- Makes project team aware of project requirements

- Clarifies legal and regulatory requirements

- Facilitates across-the-board collaboration

- Enhances response to the triple constraints of time, cost, and quality

- Provides justification for contract changes

Changes are a fact of project management. There cannot be a workable permanent boilerplate statement in a contract. This is particularly true in science, technology, and engineering projects where project execution may be subject to frequent dynamic developments. Some of the possible reasons for contract changes include the following:

- Unrealistic performance expectations
- Lack of specificity in the contract
- Lack of measurable basis for work
- Work not aligned with core business goals
- Misinterpretation of contract
- Excessive inspection of work
- Knowledge uncertainties
- Changes in delivery schedule
- Improperly executed contract options
- Proprietary and nondisclosure disagreements
- Technological advances
- Budgetary changes

VENDOR RATING SYSTEM

A vendor must be committed to the producer; the producer must be committed to the vendor. Just as customers are expected to be involved in project success, so also should vendors be expected to be involved. Customer requirements should be relayed to vendors so that the goods and services they supply to the project will satisfy what is required to meet project requirements. Selected vendors may be certified based on their previous records of supplying high-quality products. A comprehensive program of vendor–producer commitment should hold both external vendors and internal project process jointly responsible for high-quality products, services, and results all through the project lifecycle. The importance of vendor involvement is outlined below:

- Vendor and project team have a joint understanding of project requirements

- Skepticism about a vendor's supply is removed.

- Excessive inspection of a vendor's supply is avoided.

- Cost of inspecting a vendor's supply is reduced.

- Vendors reduce their costs by reducing scrap, rework, and returns.

- Vendor morale is improved by the feeling of participation in the project's mission.

To facilitate vendor involvement, the producer may assign a liaison to work directly with the vendor in ensuring that the joint quality objectives are achieved. In some cases, the liaison will actually spend time in the vendor's plant. This physical presence helps to solidify the vendor-project relationship. Also, the technical and managerial capability of the producer can be made available to the vendor for the purpose of source quality improvement. Many large companies have arrangements whereby a team of technical staff is assigned to train and help vendors with their quality improvement efforts.

THE RATING PROCEDURE

A formal system for vendor rating can be useful in encouraging vendor involvement. Vendors that have been certified as supplying high-quality products will enjoy favorable prestige in an organization. Presented below is a simple but effective vendor rating system. The system is based on the opinion poll of a team of individuals.

Requirements

1. Form a vendor quality rating team of individuals who are familiar with project operations and the vendor's products.

2. Determine the set of vendors to be included in the rating process.

3. Inform the vendors of the rating process.

4. Each member of the rating team should participate in the rating process.

5. Each member will submit an anonymous evaluation of each vendor based on specified quality criteria.

6. Develop a weighted evaluation of the vendors to arrive at overall relative weights.

Computation Steps

1. Let T be the total points available to vendors.

2. Set $T = 100(n)$, where n = number of individuals in the rating team.

3. Rate the performance of each vendor on the basis of specified quality criteria on a scale of 0 to 100.

4. Let x_{ij} be the rating for vendor i by team member j

5. Let m = number of vendors to be rated.

6. Organize the ratings by team member j as shown below:

Rating for Vendor 1 = x_{1j}

Rating for Vendor 2 = x_{2j}

Rating for Vendor 3 = x_{3j}

Rating for Vendor m = x_{mj}

Total Rating Points (from team member j) = 100

7. Tabulate the team ratings and calculate the overall weighted score for each vendor i, using the following equation

$$w_i = \frac{1}{n}\sum_{j=1}^{n} x_{ij}$$

For the case of multiple vendors for the same item, the relative weights, w_i, may be used to determine the fraction of the total supply that should be obtained from each vendor. The fraction is calculated as follows:

$$F_i = w_i \left(\text{Size of total order}\right),$$

Where F_i is the fraction of the total supply that should be obtained from vendor i. The size of the order may be expressed in terms of monetary currency or equivalent product units.

REFERENCES

Badiru, A. B., *Industry's Guide to ISO 9000*, John Wiley & Sons, New York, 1995.

Badiru, A. B.; A. Badiru; A. Badiru, *Industrial Project Management*, Taylor & Francis Group, CRC Press, London, UK, Boca Raton, FL, 2007.

Chopra, S. and P. Meindl, *Supply Chain Management: Strategy, Planning, and Operation*, 2nd Edition, Pearson Prentice-Hall, Upper Saddle River, New Jersey, 2004.

PMI, *A Guide to the Project Management Body of Knowledge (PMBOK Guide)*, 6th Edition, Project Management Institute, Philadelphia PA, 2017.

Index

Milton Keynes UK
Ingram Content Group UK Ltd.
UKHW031137141024
449569UK00006B/130